DATE DUE			

Rice and Man

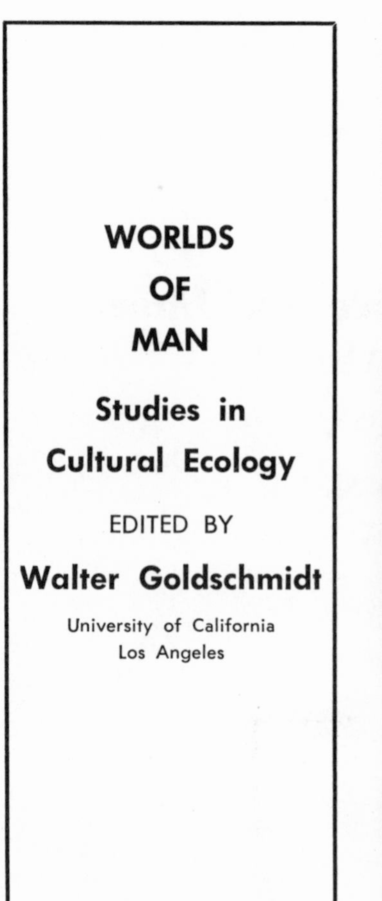

WORLDS OF MAN

Studies in Cultural Ecology

EDITED BY

Walter Goldschmidt

University of California
Los Angeles

Rice and Man

Agricultural Ecology in Southeast Asia

LUCIEN M. HANKS

CORNELL UNIVERSITY

ALDINE • ATHERTON | CHICAGO • NEW YORK

The Author

Lucien M. Hanks has taught at the University of Illinois, Bennington College, the University of California, and the University of Vermont. He is the author of numerous articles in anthropology and is currently Senior Research Associate in the Southeast Asia Program, Cornell University.

First published 1972 by
Aldine • Atherton, Inc.
529 South Wabash Avenue
Chicago, Illinois 60605

ISBN *0–202–01107–0, cloth; 0–202–01108–9, paper*
Library of Congress Catalog Number 78–169512

Designed by John Goetz
Printed in the United States of America

Foreword

To the Thai as to many tribal peoples, the basic idea underlying the modern concept of ecology seems a self-evident truth, if voiced in a different idiom. The people of Thailand recognize a unity of all life and the interdependence of man with other living creatures as part of their religious philosophy as well as in the pragmatic aspects of their daily activities. In *Rice and Man,* Lucien Hanks shows us this ecological relationship, not only in terms that have relevance for modern cultural ecology, but also in terms of the meaning this relationship has for the villagers of Bang Chan, a community near Bangkok.

Hanks' treatment of man and rice is reminiscent of those Renaissance painters who portray the details of the interior scene while, through the window, they show the broader world beyond. He depicts in detail the economic activities and social relationships of Bang Chan and places this in the context of the history and sociology of rice cultivation. Thus we see how the diverse ways by which man has cultivated rice relate to the environmental and historical circumstances and how, in turn, they shape the institutions of the cultivators themselves.

The story of Bang Chan itself leads to surprises. This village, lying but 22 miles from Bangkok—indeed, now absorbed into the greater Bangkok industrial area—was pioneered by its first cultivators scarcely more than a century ago. When one now flies across the delta of the Chao Phraya River and sees the broad expanse of padi, one is apt to think of this as stretching as far back in time as it does in the directions of the compass. Yet it was only when King Rama III pushed a canal through the region for military purposes some time before 1850 that the first farmers could enter this land, which had up to then been inhabited by roving bands of Khmer hunters and shifting cultivators. Hanks shows the successive development in the technique of rice cultivation that took place in response to the gradual filling up of the land and the altered patterns in communication and markets and the manner in which these techniques involve the collaborative efforts of family members and between families. One of his surprising conclusions, derived from a careful analysis of input and output, is that the more intensive forms of cultivation are not more profitable in an absolute sense, though, of course, they render more yield for each acre of land.

While the central theme of *Rice and Man* is the ecological framework that shapes the everyday life of the people of Bang Chan, Hanks does not lose sight of the poetics of Thai life. It is, indeed, his capacity to juxtapose the mundane and pragmatic with the poetry and magic of peasant existence that gives his work its peculiar charm and unusual value. This duality is particularly appropriate to the subjects of the study, for surely the Thai peasants themselves never lose sight of these seemingly disparate facets of their own world.

WALTER GOLDSCHMIDT

Acknowledgments

While in the main writing is a lonely job, its repercussions rock the house enough so that Jane, who read, wrote, and revised, is not listed in this book as coauthor only by force of her wishes rather than mine. I am ever mindful of Cornell University colleagues, among them Lauriston and Ruth Sharp and their graduate students of that day, who introduced Jane and me to Bang Chan and shared their lively impressions. Above all, I do not forget our instructors, the people of Bang Chan. May they find few mistakes in this recitation!

Come, Spirits of my Ancestors,
And accept this offering!

—from a Lahu prayer

Contents

Part I

BANG CHAN AND THE CULTIVATION OF RICE

In the face of chronic pressures to increase the productivity of tropical agriculture, this study of the societal concomitants of "improving" techniques derives pertinence. We seek to find out what happens to rice growers when they increase their output of rice. The physical-chemical requisites and modes of cultivation concern us less for their own sakes than as they relate to the human side of cultivation, the way of living of the growers. Indeed, these agricultural topics have been described authoritatively by such authors as D. H. Grist in his book *Rice* (London, 1959) and Takane Matsuo in *Rice and Rice Culture in Japan* (Tokyo, 1961), to mention only two important contributions in an awesomely vast literature. Our emphasis falls on the human beings who perform these tasks, knowing that every task requires at least an agent to carry it out and often far-reaching reinforcements in households and villages. Plowing and harrowing of fields require not only familiarity with soil textures and skill in manipulating implements but also a considerable knowledge, especially in Southeast Asia, of husbanding water buffalo. Land must be available for pasturing them, buildings to stable them, people to tend them. Grass must be gathered for their feed, and smoky fires lighted to keep away the mosquitoes. Thus we can begin to assess the extent of the social and psychological changes that any new agricultural technique implies. If agricultural yields must be increased, we shall know more of the human costs.

By applying ecological concepts to this scene of rice cultivation, we bring together into a whole the human as well as the plant and animal inhabitants of an area, the techniques of cultivating rice, and the physical as well as cultural environments. This book seeks to unfold the various facets of what we shall call the ecosystems of rice by focusing particularly but not exclusively on a single area of rice cultivation in Thailand called Bang Chan. Our exposition begins with a general description of Southeast Asia as a rice-growing area. After a brief statement about the botanical and historical aspects of rice, we turn to the physical environment and modes of cultivation. There follows an assessment of the work requirements for operating three typical modes of cultivation. In the second half of the book we shall describe the social and economic changes that accompanied a succession of agricultural techniques. There, over the course of a little more than a century, three modes of cultivation have been used. In successive chapters we portray the characteristic social organization that fits a particular mode of cultivation and also look to the factors that have brought about change in technique. From these materials we seek to find three distinct types of ecological relationships of rice cultivators to their environment and observe some of the factors involved in the transitions from one set of relations to another.

One value of this ecological approach lies in the new perspectives offered on certain old problems. For instance, the dyked fields are customarily regarded as important stages in the development of riziculture. Our data, however, point not only to mounding dykes in level fields but also to leveling dykes to resume another mode of cultivation. Rather than a stage in development, these dyked fields become adaptations to an environment with a moderately dense population, limited area for cultivation, and the market incentives for increasing production. To be sure these modes of cultivation may well have originated in sequential fashion, yet their contribution to human society bears no necessary relation to this sequence, for the technically final and most complex may be discarded for the simplest, when conditions are right. Thus technological evolution may well proceed in different directions and at different rates than social evolution. These and certain other implications of the ecological viewpoint for man and society are drawn in the final chapter.

Chapter 1

SOUTHEAST ASIA AND BANG CHAN

Geologically speaking, mainland Southeast Asia was part of an extensive plateau based on granitic and metamorphic rocks that seems to have covered the present Indian peninsula as well as much of the Indonesian archipelago (see Map 1). During the Mesozoic era this land mass was folded into the mountains that divide Laos and Cambodia from Vietnam, the east-west ranges from Vietnam through Laos to Burma, and the north-south spur that sets the boundary of Burma with Thailand. A continuation of the latter range forms the backbone of the Malay Peninsula. To these old ones, new mountains were added by geologically recent but titanic upheavals that formed the Himalayas, raised the Tibetan plateau, and may have sunk the great, dry Tarim basin in western China. This convulsion seems to have tilted the northern edges of the older plateau and mountain block, so that the melting snows from the new mountains ran southward. Then the Salween, Mekong, and Red rivers began to flow, though perhaps the better known of those established at this time is the Yangtze moving east through China. In Southeast Asia the three longer rivers take their places beside the shorter but older Irrawaddy and Chao Phraya, which start from the nearer and older hills.

Over the years the silts from these five rivers have settled into plains near their mouths. Unlike the sprawling Irrawaddy delta, the Salween plains are small because of the small volume of river

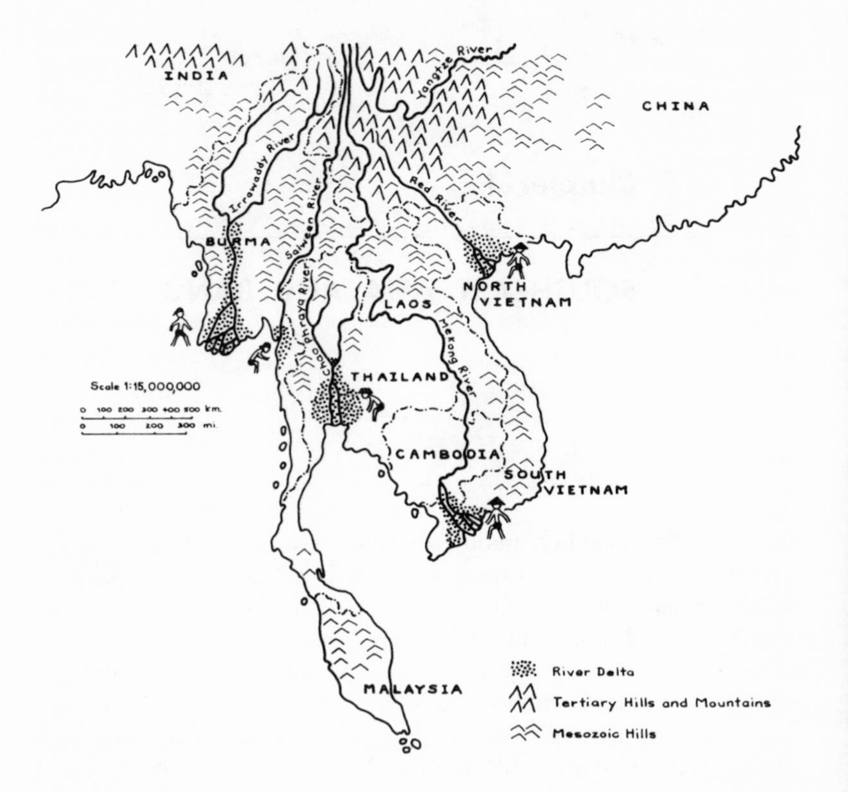

Map 1. Mainland Southeast Asia: Mountains, Rivers, Hills, and Deltas

water and minimal sediment acquired en route through rocky valleys with well-forested hills. The Mekong's enormous volume of water and heavy bulk of sediment adds rapidly to the plain at its mouth, though we have no figures on its rate of land formation. In Vietnam the Red River advances the land approximately 330 feet per year, or about the same rate as the Mississippi River. The Irrawaddy adds to its delta 165 feet per year, the Chao Phraya 15 to 20 feet each year, the Salween a good deal less. These figures refer to the point or two where advance is most rapid; the perimeter of an entire delta moves much more slowly. The central plain of the Chao Phraya is said to extend as a whole three feet a year into the Gulf of Thailand. Of course, the speed of land formation depends not only on the volume of sediment borne by the river but also, among other factors, on the depth of the ocean and the speed of ocean currents. In this part of the world where the continental shelf offered a land bridge to Indonesia during advances of the Pleistocene glaciers, the seas are shallow, the currents sluggish. Thus four of the five rivers have produced alluvial plains in their moment of extension.

All five of these plains enjoy a similar climate, average temperatures in winter months varying between 60° and 75° F. and in summer temperature standing about 85° F. The relatively cool, dry winter months are succeeded by a hot season, then by warm, wet summers with 40 to 80 inches of rain. Then the rivers rise and spread their waters over the plains, gradually receding as the rainy season ends.

On each of these plains live the greatest concentrations of people relative to the country in which it is located, for national boundaries as well as mountain chains retard the movement of people. In Burma densities reach their highest at 206 persons per square mile on the Irrawaddy plains. Thailand's central plain along the Chao Phraya achieves 520 persons per square mile. Still more heavily populated Vietnam mounts to top densities of 520 per square mile. In comparison, only the most urbanized states of America—New Jersey, Massachusetts, and Rhode Island, for example—reach and exceed the latter level of density, whereas the vast majority of states have densities of less than 100 per square mile.

All of these areas have long been dedicated to the production of rice. According to the *Trade Year Book* for 1967, a publication of the United Nations' Food and Agriculture Organization, the total world exports of rice have varied in the decade 1956 to 1966 between 69 to 84 million short tons of 2,000 lbs. Of this, two-thirds to three-quarters come from Asia, with North America (mainly the United States) and Africa (mainly the Malagasy Republic) making up the bulk of the remainder. Burma and Thailand together produce for export about 32 million short tons or approximately 43 per cent of the world export crop. Such figures reveal little about domestic production, but if we may conservatively assume that these two countries export half of the rice they produce, then together they produce nearly as much as the total world export. In America the comparable grain is not rice but wheat, yet even the total wheat crop during any single year of the 1960s never exceeded 40 million short tons.

From these broad considerations, let us turn to Bang Chan, a microscopic dot on the central plains, the valley of the Chao Phraya River in Thailand (see Map 2). We are drawn to it not because of any special productivity, since rice yields there are actu-

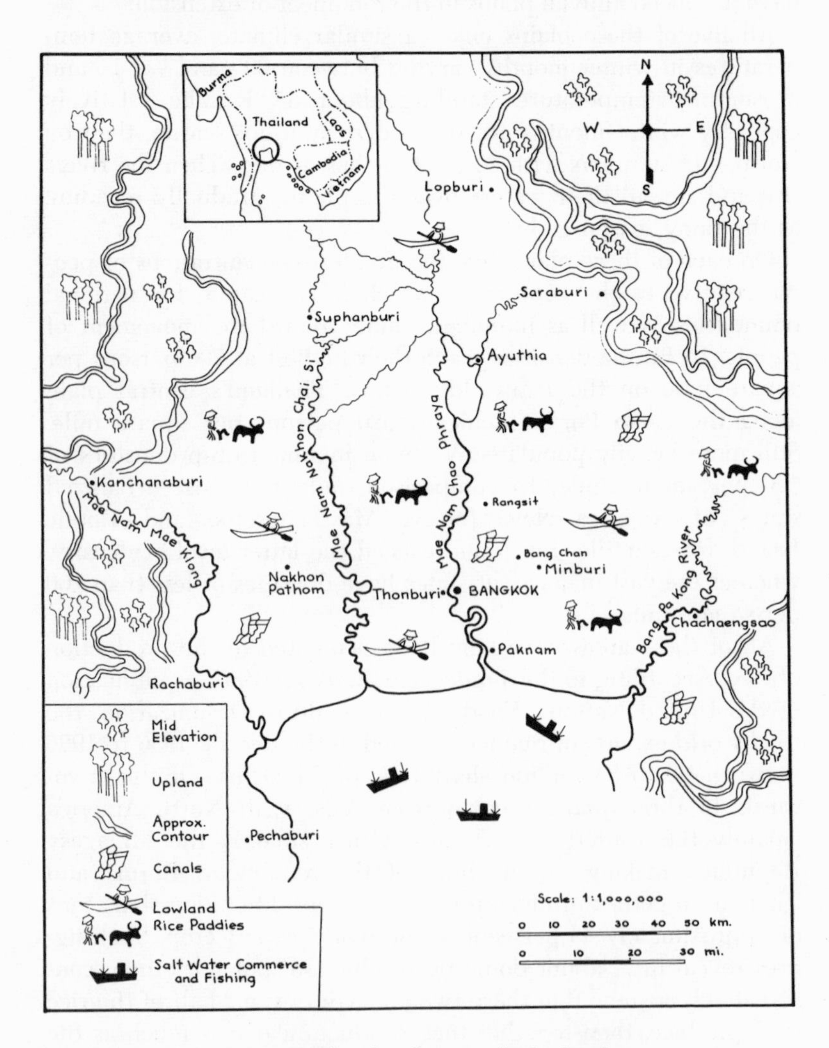

Map 2. Sketch of Central Thailand

ally lower than those in northern Thailand or Vietnam. Rather, Bang Chan offers a choice setting to observe the relations between people and their environment through the successive techniques used to cultivate rice. There one can see compressed into a maximum of 120 years the events that elsewhere may have required 3,000 or more. Rice was raised for centuries before dyked fields were employed, but in Bang Chan dyked fields were introduced in 1935, just about 85 years after the founders of Bang Chan began with the simplest tools. Indeed, the very recency of this event adds an intimacy of observation that evaporates when dealing with the historically remote.

The Chao Phraya plain where Bang Chan is situated approximates a right triangle with a 92-mile base on the Gulf of Thailand. Its altitude roughly parallels for 152 miles the mountains bordering on Burma as it moves northward to the confluence of the Mae Ping and the Chao Phraya rivers. The hypotenuse returns southeasterly along the base of a series of hills to the northeastern corner of the Gulf of Thailand. Though this plain extends geologically another 155 miles northward, the rough contour and the fact that water is not easily available make this northern part a secondary rice-growing area at best. The southern part shown in Map 2 enjoys the luxury of well-dammed and channeled water from the Mae Klong, Chao Phraya, and Bang Pakong rivers. Here live the densest concentrations of people who produce more than half of the rice of Thailand on slightly less than half the lands devoted to rice production.

In this southern part of the central plain, Bang Chan lies east of the Chao Phraya and about 22 miles by highway from Bangkok (see Map 3). Bang Chan Canal, lying in the center of Bang Chan, is a tributary of the large Saen Saeb Canal, the old avenue to Bangkok. The land is as flat as a calm sea, but the many canals, dyked fields, and house mounds sharpen its contours. The big canals like Saen Saeb attain breadths of 40 feet and depths (according to season) up to 7.5 feet; small ones, a bare yard in width, are often impassable by boat during the dry season. The dykes around the paddies and the house mounds rise or fall on contour waves varying less than a yard in 1.2 to 1.8 miles. These barely perceptible elevations become important for

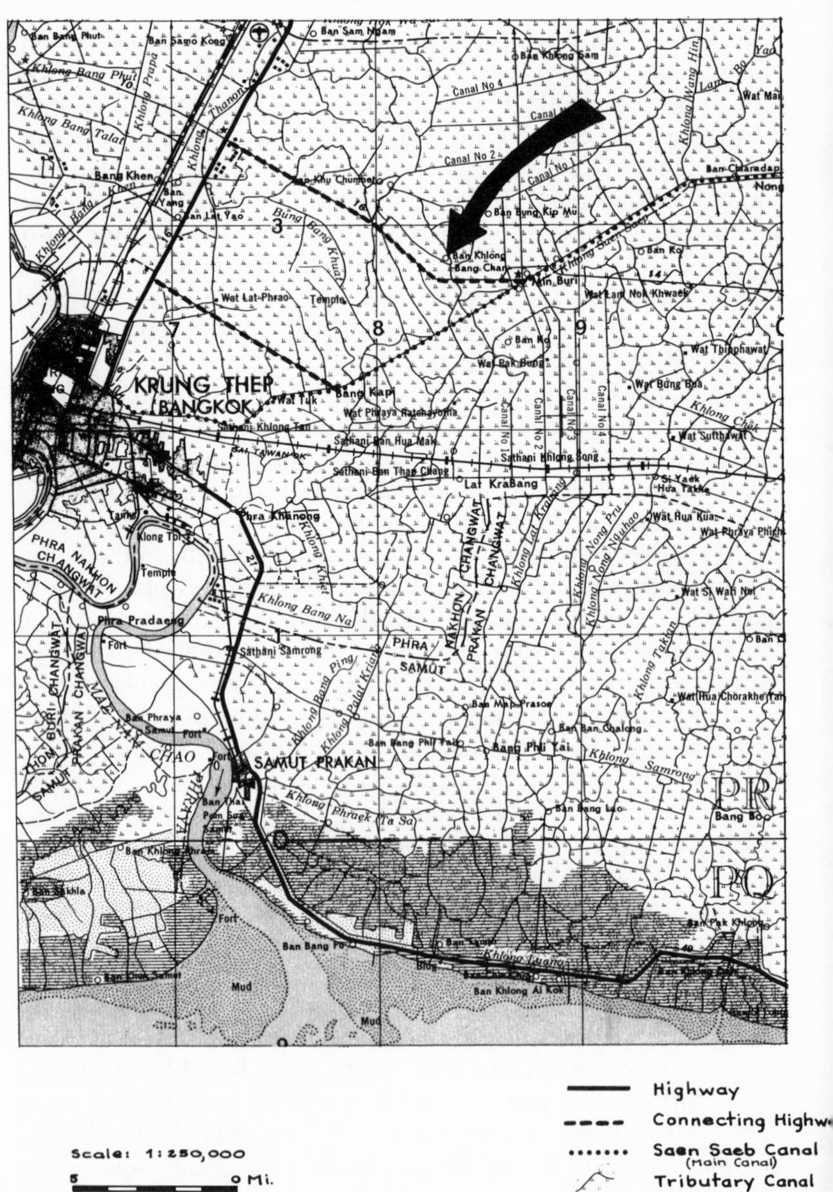

Highway

Connecting Highway

Saen Saeb Canal (Main Canal)

Tributary Canal

Rice Fields

Mangrove Swamp

Provincial Border

Scale: 1:250,000

5 0 Mi.

5 0 Km.

Courtesy of the U.S. Army Map Service

Map 3. Bang Chan and its Environs

determining the time and depth of water flowing into a field, for in this country with skies spanning the horizons, an elevation of a yard may delay planting time by a fortnight or more.

In January water runs off the fields into the ebbing canals. Through the next rainless months and particularly during the heats of April and May, the shallower canals turn into weedy courses flanked with dyked fields of clay baked to bricklike hardness. Then in June the waters begin to return, at first not because of local but northern rains that filter into the irrigation network. July signals the start of plowing with sudden heavy local downpours that, mounting in frequency through the succeeding months, produce the general flooding of September. November is a breathless month with the rains terminating in humid weather and the canal waters noticeably ebbing with the approach of the New Year. The Thai divide the year into three seasons: the cool of December through February, the hot season of March through May, and the rainy season that begins in June.

Bang Chan is only about as old as the State of Kansas, for its first settlers appeared during the early 1850s. It has been continuously settled since at least 1861, the year Kansas became a state. Because Cornell University's Thailand project concentrated its observations during the decade 1948 to 1958, we shall introduce the area as seen during this period.

In 1953 Bang Chan's population lived as 298 households, of which 84 per cent derived a living mainly from agriculture. The remainder worked as storekeepers, teachers, priests, peddlers, plus one man who worked in the district office at Minburi. The agricultural section of the population consisted in 54 per cent who owned and operated at least a portion of their lands, 31 per cent farm renters, and 13 per cent farm laborers. Among the landowners about two-thirds rented some portion of the land they worked in order to extend their acreages. Except for some 5 per cent used for house and store sites, all land was given over to cultivation.

In all, there were 1,694 people living in an area of 2,600 acres. This furnishes a population density of 422 persons per square mile, one of the higher rural densities for Thailand at that time. In this population, children under 15 represent 38 per cent of the population, a figure near the average of other samples in Thailand. In the U.S. this age group contributed 31 per cent to the population in 1960, somewhat more than Scandinavia,

England, and Italy's 25 per cent or less. Those 60 years old and over formed 7 per cent of the Bang Chan population, as compared with 12 per cent in the U.S. and 15 per cent or more in many Western European countries. Females outnumbered males very slightly. As for births, Bang Chan's crude rate was 26.2 per thousand while the comparable figure for the U.S. in the 1960s was 22.4, a figure that had risen from about 17 during the 1930s. Though no record was made of deaths in Bang Chan, the median age of the population lay somewhat above 20 and suggests an average life expectancy of 25 years or less, a figure even below that of the poor in Europe a century ago. Today in the U.S., life expectancy stands in the sixties. Clearly, Bang Chan turns over its population rapidly. Lacking systematic studies of deaths in the locality, we cannot do better than to recall the generality that potentially lethal intestinal diseases such as typhoid fever and dysentery are common in tropical countries. Clinical examination of Bang Chan schoolchildren revealed a high incidence of intestinal parasites. Malaria attacks occur, but Bang Chan does not consider them grave. Cholera has ravaged the area two or three times within the present century and is greatly feared. With an approximate food intake of 2,600 calories per day, nutritional deficiences present no special problem, although animal proteins fall significantly below American levels of adequacy.

Bang Chan extends along forking canals lined with unevenly spaced residences, each in its own grove (see Map 4). Climb into one of the sampans with almost no freeboard and less stability than a canoe; despite its precariousness, this is the best way to visit the area. Soon there passes a long dugout type of boat, loaded with a metal-tipped plow, and paddled vigorously by farm boys on their way to some distant field. A middle-aged woman, her straw hat shaped like a great inverted basin, moves her boat steadily and slowly to the market with a cargo of lotus, duck eggs, and squash. A largish undecked ark, schoolbound and bristling with young paddlers in their blue and white uniforms. careens a little unsteadily, almost colliding with the tiny skiff of a mustachioed old man. A ten-year-old girl steers from its stern.

Some of the dwellings enjoy a broad verandah, some are raised a little higher above the ground than others and require a longer ladder to reach them, some are of the blunt-roofed new style and others feature the peaked roof of the old style. Here

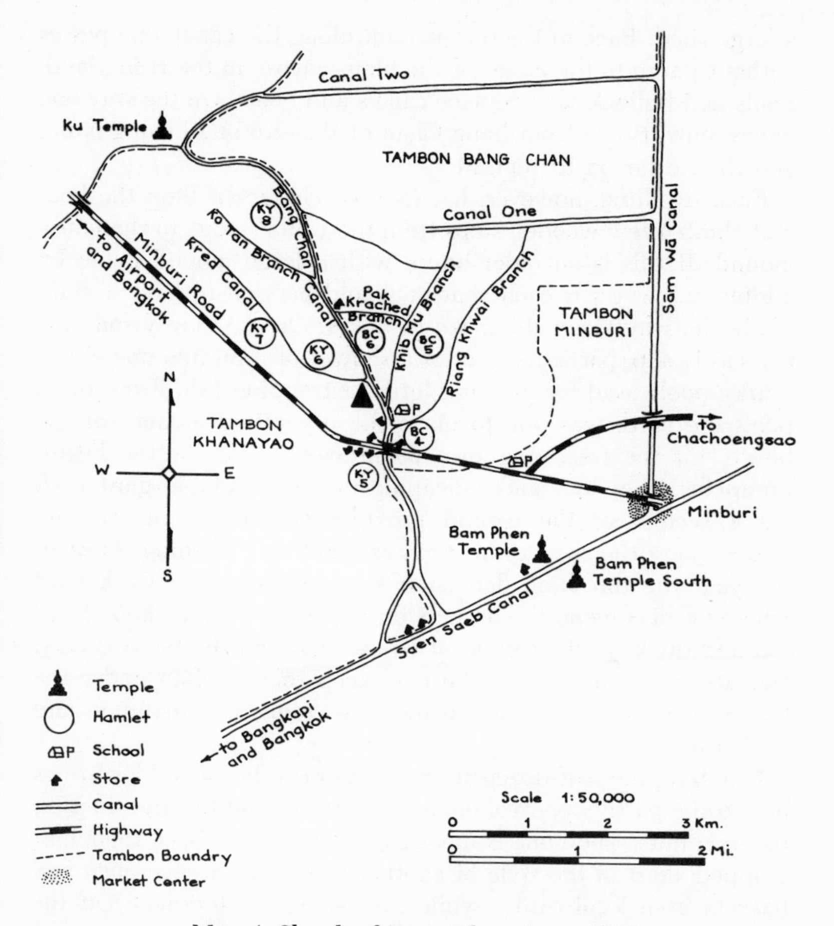

Map 4. Sketch of Bang Chan Area, 1957

and there are small bamboo houses at ground level with thatch instead of tile or corrugated-iron roofs; they are recently built, unprotected by trees, and stand on earth mounds unadorned by vegetation. A few boats may be tied in front of an open-fronted hut, the site of a local store selling a few rusting mattocks and pink lime for betel nut chewers. The main focus of the scene rests at the somewhat battered temple compound, a few glistening spires above a collection of weathered wooden buildings. There one may easily pass a few hours talking to the resident monks or, attracted by the singsong of recitation, visit the school housed in a large shed. Back in the sampan out along the canal, one passes without pause to the edges of the plain where, in the rising land, roads and bullock carts replace canals and boats. On the way one moves unwittingly from Bang Chan to Wat Ku or Khlaung Nung and thence far off to Rangsit.

Each dwelling, however, has its own character. Stop the boat and climb a few wooden steps from the water's edge to the house mound. If this is an older house with a large mound made by adding a few yards each year, it would be well to have a child guide the way along the maze of dykes. On the clayey soil, one can easily slip, particularly after a shower of rain, into one of the murky pools used for growing lotus or trapping fish. From these pools the earth was dug to make the dry raised mound for the house, for the trees that provide firewood, and for the beans, pumpkins, eggplant, and pineapple in the vegetable garden. A major section of the mound provides the important working space where the sheaves of grain are laid for threshing. Most of the year the threshing floor stands vacant, a sunny work yard where farmers mend the harness for the water buffalo, and chickens and ducks peck at grain or straws lying on the tamped clay. But at harvest time, some farmers keep their buffalo under the house in the forest of supporting posts, tools of cultivation, and cool shadows.

Climbing the half-dozen steps on the main house ladder brings one to the partly open verandah, a feature of all but the simplest bamboo huts. Then one is among aging grandmothers, their hair cropped short in the style of another time, who weave mats and baskets from local rushes while overseeing the toddlers. Off the verandah, in an alcove sheltered from wind and rain, a younger woman squats on her heels before an open fire cooking the fish and vegetables that garnish the rice. If the house is of the old style, one's eye may pass over the teak-paneled siding and climb over the shingles to the peaked roof with its pointed projection at both gable endings. Here is the ungilded, homemade simulation of the cosmic serpent, the naga, that adorns every important temple. It seems to descend from the heavens and slide gracefully down the edge of the roof to the eaves where its raised head is indicated by a soft reverse curve. With four such guardians, one at each corner, the house is secure from malign influences.

The main room of the house is entered through a door off the verandah. The room is quite bare, lighted by three or four shuttered windows; here the household members sit on the floor to eat their meals and unroll their mats to sleep. In one corner a small dusty table supports a pot of bristling incense sticks that once burned before the colored picture of the Lord Buddha hanging on the wall. Add to this a framed picture of the king, a fading photograph of a family reunion, and a cabinet or two for storage of mats, clothing, and blankets, and one has pretty well exhausted the contents of the room. Some houses contain a small extra inside room shut off by added teak paneling. Here a newly married daughter and her spouse might live for a year or two, or an aged grandmother; subsequently it would be used for storage.

How should we understand this scene of "ribbon settlements" where houses line the canals, each one on its own field, and extend indefinitely across the plain? The temple, school, and store alone stand out on the landscape, but this is scarcely unique. As recently as 1930 one might find rural sections of the American Midwest with a church, a school, and a general store the sole breaks along roads leading monotonously past fenced fields with their barns and farmhouses. When school and store, or store and church, were clustered together, perhaps at the crossroads, people gave it a name like Cottage Grove or Pheasant Branch, but no signs announced the fact. One might pass several such centers not knowing their names. So parts of rural America shared with Bang Chan continuous settlement of people living on the lands they worked—named areas with religious, educational, and shopping facilities for the nearby residents, and unproclaimed boundaries. The main difference lies in the size of the land worked. Instead of the traditional Midwestern quarter section (160 acres), Bang Chan people worked 4 to 40 acres.

As in America, the Thai rural areas have their larger market centers. The nearest to Bang Chan is Minburi, formerly reached by boat by descending the canal and turning east on Saen Saeb Canal. The new road completed in the early fifties and bearing occasional buses can take a man there in ten minutes from the bridge crossing Bang Chan Canal. In Minburi one can buy hardware or cloth from one of two dozen stores run by Chinese proprietors, take a sack of grain to one of the five towering steam rice mills, eat a bowl of noodles in a restaurant, visit the local government offices, and on Saturdays attend a cockfight. Some Bang Chan people went there on the average about once a month, but others preferred to take the bus in the other direction along the highway to Bangkok. In an hour they would be at the Pratunam market. A few old and young had visited neither place. Other market centers like Minburi lie dotted through the plain at distances of approximately 25 miles apart; Bang Chan resembles rural Iowa in this respect.

Despite important differences in the character of local government, the governing of Bang Chan resembles government in rural America in a few respects. Thailand's 71 provinces subdivide into three levels: district, commune, and hamlet, roughly comparable to American county, township, and village government. The Thai equivalent of the county government in Minburi is conducted by the district officer, an official appointed by the Ministry of the Interior in Bangkok. The local court, police, jail, schools, health, and agricultural services operate under his jurisdiction. At the lowest level, the hamlet headmen are in effect nominated by the locality and appointed by the district officer. They report local needs to the district officer and after monthly meetings relay certain announcements from him to the hamlet. The intermediate level, the commune, is under a hamlet headman nominated by his fellow hamlet headmen to be commune headman; he is also appointed by the district officer. He enjoys a slightly higher salary than hamlet headmen and in many areas helps to gather taxes. With 35 households it is ordinarily possible to form a hamlet, and with ten hamlets a commune. Thai law, however, avoids precision on these points.

The demarcation of governmental units in Bang Chan shows administration at its most arbitrary. It was easy enough about the year 1900 for the planning committee of the Ministry of the Interior to draw the provincial boundary that placed Bang Chan in the same province as Bangkok. When it came to designating the districts in an all-but-featureless landscape, they readily seized upon Bang Chan Canal to set the boundaries between districts. So on the east side of Bang Chan Canal the inhabitants are residents of the District of Minburi, commune Bang Chan; on the west, residents of the District of Bangkapi, commune Khanajaw. The area under study called Bang Chan consists of three hamlets in commune Bang Chan, four hamlets in commune Khanajaw (see Map 4).

In what respects is Bang Chan more than an arbitrary slice of a countryside? No easy way presents itself to delimit the area without cutting off some more or less vital transactions. Our investigations sometimes carried us beyond the seven hamlets of communes Bang Chan and Khanayaw. Nevertheless, the local residents of these hamlets tend to send their children to the school at Bang Chan, send their sons to enter the monastic order at Bang Chan temple, and buy certain goods at the stores of the area. The man who lives a little nearer Wat Ku than to Bang Chan's temple is apt to send his children there to school and visit the temple there. But no firm rule forbids sending his children to a more distant teacher or attending a more distant temple ceremony. Bang Chan with its school, temple, and stores, rather than a discrete segment of society, becomes a center of orientation for the rice growers in its vicinity, much like Cottage Grove, Pheasant Branch, or some other corner of rural America. Moreover, we can imagine a rural area with just this amount of cohesion arising on an American prairie before counties and townships emerged. Then when officials drew subdivisions on a map, they ran the lines up the center of a road, leaving the store in one county, the church and school in another. But the people, whether in the Midwest or in Bang Chan, step daily across the lines without concern for political boundaries.

With this brief glimpse we shall part from Bang Chan, to observe in Chapter 2 something about rice itself, and, in Chapter 3, the various environments in which it grows and is grown. Then, having examined in Chapter 4 the particular human contributions to rice production, we shall return in subsequent chapters to Bang Chan itself for a second look, following the sequence of changes of environment and society which have molded this area to its present shape.

Chapter 2

RICE AS PLANT, CROP, HISTORY, AND WORLD OUTLOOK

Botanists tell us that rice, along with wheat, oats, barley, sorghum, Job's Tears, maize, and bamboo, as well as many other less well-known plants, belongs to the extensive family of grasses or *graminea.* All have in common stems with nodes, for which bamboo offers a familiar example. The long, thin leaves of grasses have veins running parallel, and this distinguishes them from many other plants with branched veins in the leaves, such as oak or maple. All have varying shapes of inflorescence: here we refer to the headed clusters of minute flowers which form the tassels of maize, the busy tops of timothy, or the bearded seed-bearing cluster of barley. Rice differs from such cousins as maize not only in size, but also because its fertile seeds grow from the inflorescence at the top of the plant, instead of at the side where the ears of corn grow. The inflorescence or "panicle," as this particular shape is called in rice, tends to be beardless and is always fan-shaped. At harvest time this laden head has five or more slender radiating ribs to which the grains have been attached like rows of single, pendant jewels along the length of each rib. The grains themselves resemble those of wheat or oats, the edible fruit at the center being covered with a brown husk. A good shaking suffices to loosen the grain from the panicle, and by pounding it in a mortar or pressing it between rollers, the kernel pops from the husk. In Southeast Asia, women winnow the chaff from the grains by tossing a trayful from the mortar and letting the wind blow the grains clean of husks and dust. The grain is usually cooked by boiling or steaming.

Enthusiastic taxonomists of the eighteenth century gave the genus of rice the Latin name *Oryza,* and their nineteenth century confreres found many specimens with enough morphological peculiarities for separate species. Thus the genus *Oryza* expanded with dozens of species. In compendious botanical dictionaries, each author argued for the merits of his particular mode of classification. Around the turn of the century, a new generation of botanists became interested in plant physiology and discovered that by varying conditions of growth they could alter structural features of plants. Repellent thorns disappeared when desert plants grew in moist, nutritious soil. Dwarfed alpine species doubled their heights in the warmth of a valley. Thus many a genus and species named by the older generation had to be reclassified on the basis of more persistent characteristics. From dozens of species of the genus *Oryza* have emerged some 25 that, like members of the ubiquitous grass family, grow on all the world's continents.

The greatest number of species occurs in Asia and Africa, though Australia and the Americas also have unique species, if we may use the Japanese system of taxonomy. However, the latest edition of Gray's *Manual of Botany* (8th ed., 1950) lists five North American species of a genus called *Oryzopsis,* meaning "like rice," and none of the genus *Oryza.* As for "wild rice" gathered by Indians in the marshy edges of Minnesota lakes, all agree in assigning it to a third genus, *Zinzania,* thus dubbing it a near cousin of domesticated rice known botanically as *Oryza sativa.* But many questions of classification remain unsettled.

"True rice," *Oryza sativa,* is an amazingly adaptable plant. It grows like wheat on dry slopes as well as in deep pools of water. Japanese farmers cultivate it as far north as Hokkaido, so that on the basis of latitude it might also be grown in southern New England. We have found upland rice fields in Thailand at altitudes approaching 4,500 feet above sea level as well as in the brackish tidal flats of the Gulf of Siam. In the Himalayas rice is said to grow at altitudes of 10,000 feet above sea level.

These variants produce no single type of seed. Some plants yield long, narrow grains, while others give oval shapes. The rice we know is dry and easily separated when properly cooked, but the glutinous type remains sticky. If we turn to growing conditions, some varieties thrive on light, sandy soils, but will produce little in heavy clay. There are the rices that mature in 90 days and those that require 150 to 260 days to develop grain. Certain varieties develop inflorescences when the hours of sunlight begin to diminish, though their photoinsensitive stepbrothers patiently wait out the classic internal sequence of development. So we find thousands of varieties of the single species *Oryza sativa*, each with its own virtue.

Again the painstaking work of Japanese botanists has helped order the confusion by distinguishing two subspecies: *Indica* and *Japonica*. Like all of the species, both have 12 chromosomes, or multiples of 12 up to 48. The *Indica* types have long-grained rice, tend to develop on the basis of an inner timetable, and are responsive to temperature. In contrast, the *Japonica* varieties have oval-shaped grains, mature in response to changes in duration of sunlight, and continue to grow despite fluctuations of temperature. With certain exceptions the *Indica* varieties occur on the tropical mainland from southern China to India, while the *Japonica* varieties are found in Indonesia and Japan. In addition, the hybrids of the two subspecies tend to be sterile, indicating a genetic differentiation that may have developed during the centuries of agriculture.

We can conceive how this differentiation operated. As rice cultivation spread, each new field with its peculiar qualities of light, moisture, temperature, and soil set the conditions for advantageous mutations and directions of variation. Year after year each locality of cultivators selected the handsomest, the tastiest, and the most sweetly perfumed to plant in the coming year. Where settlement or strain of seed remained stable, there developed the special virtues that characterize each variety, bearing the scars of drought and epidemic, the shape and color that please.

So the genetic studies of rice varieties associated with special localities suggest a history of rice to the botanists. The clue rests on the principle that those varieties are more akin when crossing produces fertile progeny; when the progeny is sterile, or sterile in a greater percentage of the cases, the two crossed varieties are more remote. In this manner they discovered that varieties from south China are more akin to Japanese and Indian varieties than either of the latter are to each other. They thus inferred the ancestral varieties to exist in China, and, indeed, the studies of other crosses also point to southern China and Southeast Asia as the ancestral home of *Oryza sativa*.

To what extent does archeology substantiate these inferences offered by the botanists? In India the earliest traces of rice date no earlier than the first millenium B.C. Prior to this time the populations of northern India, like those of the Tigris-Euphrates city states, lived chiefly on barley, wheat, and millet. In Japan, only the final stages of Jomon culture suggest the addition of rice cultivation to Jomon hunting and fishing; this may well indicate that the culture borrowed this practice from the newly arrived Yayoi people, about 300 B.C. at the earliest. In the Pacific islands, rice came some time after the Melanesians and Polynesians had departed from the mainland, for yams and taro are their principal crops, though recently introduced rice grows healthily today in Hawaii as well as in parts of Micronesia. Rice culture in offshore islands like Taiwan and the Philippines, as well as Indonesia, seems to have arrived as a later element to the scene than, say, yams and taro.

In China we learn again of the late acquisition of rice. A few sites of the Neolithic Yang Shao people of the Yellow River region reveal in the upper layers signs of rice cultivation. The deeper layers indicate that these people at an earlier period raised millet, wheat, and barley in fields they cleared by burning. Credit for being the first rice growers goes generally to the successors of this early culture, the Lung Shan people, who seem to have stabilized agriculture around permanent villages, inventing such standard tools as the sickle and the hoe. Possibly they discovered irrigation, fertilizers, and fallowing to maintain the productivity of their fields. In any case their culture spread north into Mongolia and Manchuria as well as southward to the Yangtze valley and beyond. These we may most conservatively date at 1450 B.C. with the beginning of the Shang dynasty. Certainly it would not be a significant distortion to add five hundred years and set a date for rice cultivation near 2000 B.C.

Although botanists and archeologists agree that rice cultivation originated on the east Asiatic mainland, a discrepancy of several thousand miles occurs between the location of the first cultivators in north China and the botanists' indication of its origin in the region south of the Yangtze River. Here we may appeal to the possibility of climatic change, that rice did grow in the north during a wetter and perhaps warmer post-Pleistocene period. Traces of elephant, rhinoceros, and water buffalo are found in the north, associated even with Neolithic and early Bronze remains. If rice grew then, we wonder why the cultivation of this luxuriant and nutritious grain by-passed the Yang Shao people and began only with the later Lung Shans.

Another lead draws us to the culturally simpler Hoabinhian peoples of northern Vietnam and south China, whose remains occur as far west as Szechuan. They were hunters and gatherers whose pebble implements may have been used to fashion bamboo weapons for killing deer, wild pigs, and bovines. Since the human bones associated with these finds suggest a negroid population, we are speaking of the physical type that now inhabits Melanesia, and has been all but replaced on the mainland of Asia by Mongoloids. We presume a considerable antiquity to these people, and, relevant to our interests, they may have been the first to possess rice, which they seem to have gathered wild in northern Vietnam. Thus, in this version, which is more consistent with the botanists' findings than the theory of a spread from the north, rice culture began in the south with the Hoabinhians and spread north to the able Lung Shan farmers.

Some day these hypotheses will be tested by dating techniques, but so far we have not yet even fitted the stratigraphic relations of Hoabinhian to Yang Shao or Lung Shan. The first batch of absolute dates—10,000 B.C. for ceramics in Japan, about the same date for agriculture in Southeast Asia, and 4,000 B.C. for bronze—suggests that, in this part of the world, the Neolithic Age, as well as the Bronze Age, may have preceded the Middle Eastern Neolithic from which Europe took its start.

During its many centuries of association with man, rice in raw and cooked form has assumed special meanings, much as in the Occident, bread has become "the staff of life," the food to be ceremonially "broken."

In a verse from Arthur Waley's translation of the *Shih Ching* (*The Book of Songs,* 1954), a collection from the Chou period in China (ca. 1100 B.C. to 500 B.C.), we find this reference to rice:

Abundant is the year, with much millet, much rice;
But we have tall granaries, to hold myriads,
Many myriads and millions of grain.

We make wine, make sweet liquor,
We offer it to ancestor, to ancestress,
We use it to fulfill all the rites,
To bring down blessings upon each and all.

[p. 161]

Here we may think of the unhusked rice brought in from the fields and stored in elevated bins to protect the crop from ravaging rats and mice. From this grain was distilled a strong colorless liquor to offer the ancestral spirits who protect the household. So a good crop helped keep the old alive and the young from dying too soon; it helped bring success to the hunt, victory in battle, and might attract a richly garbed bride to the household.

In many parts of Thailand today rice is deemed animate and so grows like other animate creatures. The rice mother, Mae Phosop, becomes pregnant when the rice flowers bloom, and as the grain grows, she, like any pregnant woman, delights in scented powder and bitter-tasting fruit. Her offspring, the rice, has its *khwan* (soul), like other animate beings. Along with the harvest, the soul too must be ritually gathered and taken to the storehouse. Then when the crop is sold, the buyer from the rice mill takes the raw grain, but in a handful of grain carefully returns the soul to the farmer to impregnate next year's crop.

Some people, like the northern Chinese, treat rice as bread is treated in a contemporary European household, as an addition to the main courses of soup, viands, and vegetables. In the more southerly parts of the Far East, rice is the staple heaped upon a plate, to which the sauces, condiments, and curries add flavoring. Indeed, Thai farmers sometimes refer to rice in speaking at pious moments of the present *yuga,* that long epoch of thousands of years between the forming and the dissolution of the universe. During the first quarter of the present *yuga,* when virtue dominated world affairs, man could eat rice by itself with perfect satisfaction. Now, in the middle of the *yuga,* when vice has increased, the nourishing qualities and flavor of rice have decreased. Rice alone no longer satisfies, so peppers and curries are sought to improve its flavor.

Nonetheless, its position at the heart of the meal survives. A heaping plate of rice becomes more than just the contentment of satiation or the nutrients to sustain our bodies. According to some Thai villagers, man's body itself is rice, and eating rice renews the body directly. Babies grow within their mothers' stomachs where they sit eating the food that the mother eats. Tissue is made of rice because it derives from rice. Mother's milk is blood purified to a whiteness. Just as mothers give their food and bodies to nourish children, so Mae Phosop, the Rice Mother, gives her body and soul to make the body of mankind. Thus the rice growers' image of man becomes rice itself; perhaps, according to this vision, man differs slightly from other living creatures, largely because of the diet that sustains him.

Chapter 3

THE ECOSYSTEMS OF RICE IN NATURE

Seen in an ecosystem, man is no longer the measure of all things nor the master of nature. He is bound intimately to the grain he would grow, for without his sensitive observation of wilting and flourishing, these plants might have dwindled to extinction like the passenger pigeon and buffalo grass. Without the magnificent adaptability of rice and its responsiveness to man's nurturing, certain tropical civilizations—Golconda in India, Srivijaya in Java, and the Khmer of Angkor in Cambodia—based as they were on riziculture, could never have come into being. Little wonder that the greatest horticulturalists have come from the Orient, where nature is not considered passive. Like the haiku poets who converse with a still pond, they believe that man, properly trained, develops sensitivity to the voices of plants, and plants can hear man's entreaties.

Ecosystems follow the oriental penchant for finding cycles in nature. Though we in the Occident acknowledge that man emerges from the dust and returns cyclically to it at death, more frequently our tracing leads to linearity. That package of rice so easily plucked from the shelf at the supermarket usually follows a line from the fields to the mills, then to the wholesalers and retailers, finally to our table. There it stops, yet it might be led back to the field quite directly, since man and animal transform rice into fecal matter, which, used as fertilizer, helps nourish the next crop. For rice to grow, a multitude of cycles must coincide for a moment of cosmic time. The *Yang* of spring must have emerged in the seasonal cycle; rains must fall in the round of evaporation and precipitation; man must stand at his strength between the weakness of youth and age; soils must be eroded from the mountainside between the uplifting and the decaying of hills. Without these and many other corresponding cycles the planting would fail; those past waves of tropical climate that swept over our world brought no rice crops to warm Devonian Labradors of Silurian Patagonias.

In the cyclic view of nature man is neither master nor pawn but giver and receiver. However small the plot he clears in the forest, he alters nature, and in order to sustain his plot he must work to clear it and keep it cleared. Once the labor stops, the grasses dependent on sunlight yield to bushes which in turn disappear under shading trees, for the scene depends as much and no more on man for its order than upon rain or chromosomes. Perhaps as a result of common heritage, the Orient and native Indian America have long recognized the implications of these observations, that harmony must prevail between man, heaven, and earth. The Chinese empire tottered, and Navaho Indians became feverish, from failing to respect this harmony, long known in the Orient and emphasized by the Taoists, among others. Thus James Legge (1927) translated the text *Chwang Tze:*

> There is nothing which has not its condition of right; nothing which has not its condition of allowability. But without the words of the water cup in daily use and harmonized by the Heavenly Element (in our nature), what one can continue long in the possession of these characteristics?
>
> All things are divided into their several classes and succeed to one another in the same way, though of different bodily forms. They begin and end as in an unbroken ring, though how it is they do so be not apprehended. This is what is called the Lathe of Heaven; and the Lathe of Heaven is the Heavenly Element in our nature (Part III, pp. 143–44).

So dry wells that follow draining swamps, and dust storms that follow tilling the desert's edge may be teaching man better to harmonize his acts with heaven and earth. Even today we may begin to write the rules of natural morality, which, once seen more fully, may resemble the reciprocities between neighbors invoked by Moses and Mohammed, as well as by Lao Tze.

In what follows we shall be dealing with four techniques for producing rice: gathering, shifting cultivation, broadcasting, and transplanting. Instead of considering them in an evolutionary sense, we shall regard them simply as forms of adaptation. People gather rice today as they did 15,000 years ago, including in that time span the uncertain date for the Hoabinhians. We have met cultivators who abandoned the sophisticated techniques of transplanting for shifting cultivation. No irrigated land was left for them in the valley, and so they climbed the hills to plant their fields.

With this in mind we may observe that the predominant techniques today are shifting cultivation and transplanting. Shifting cultivation is found in a broader area but transplanting is practiced by more people (see Map 5). Gathering has all but disappeared, and broadcasting is restricted today mainly to the western coastal region of Burma, a few parts of the central Plains of Thailand, and the lake section of Cambodia.

THE GATHERERS

Where swifter rivers meet the slower tides and drop their silt, land forms into a delta, the river erecting barriers that limit its course. Exposed directly to storms these mud bars are washed away, but eventually new breakwaters build up farther to seaward, and silts form behind the sheltered spots. Where mangroves cantilever over the tide, their tangled roots help hold the land. The falling leaves give organic nourishment to clays, and the day comes when the last storm has washed away these crumbs. Then soils may gather, and mangroves yield to grasses and sedges. At the mouth of the Ganges and the Amazon, species of native rices have been reported, and one might also expect them in the Nile, the Mississippi, and the Yangtze.

So for a few moments in the cycle of mountains becoming plains and washing into the sea, the silt comes to rest as a gray mud, but once dry and above the water in the tropics, it may be leached by rains into reds and yellows on its way to becoming laterite or sand. The soluble salts of the soil drain away, leaving the resistant ferric, aluminum, and manganese oxides of laterite or the silicates of sand to await in suspended animation a new washing or a future upfolding of mountains. On such spots, be-

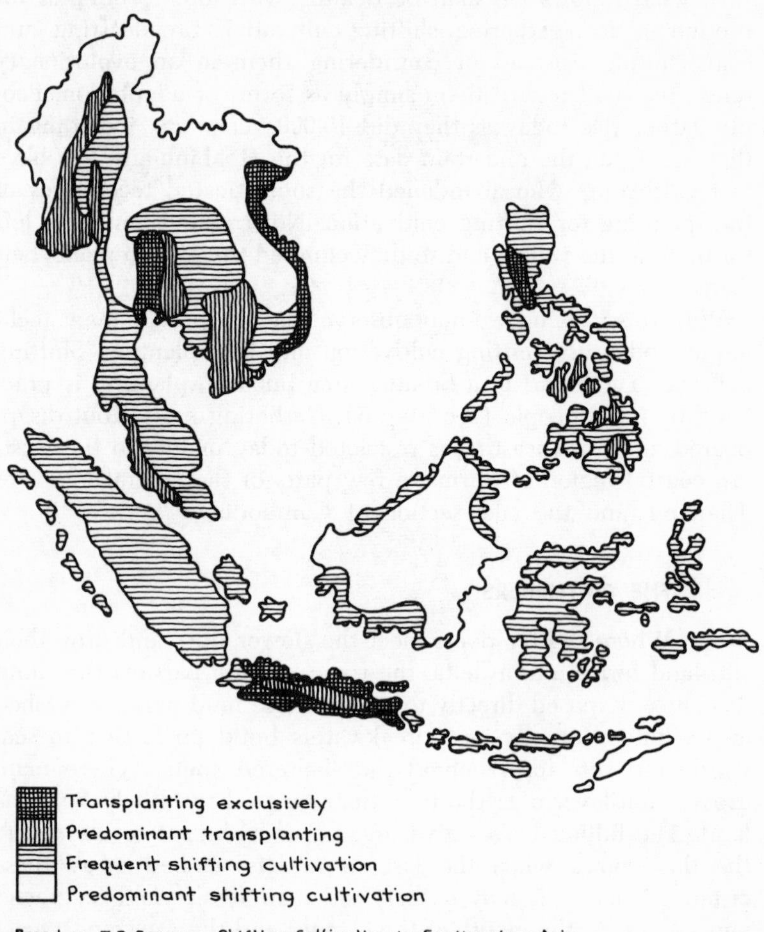

Map 5. Major Locations of Transplanting and Shifting Cultivation

fore the soil loses its nutrient salts near the surface, grasses grow for a brief moment in the floral succession between the mangroves and shaded forest. Yet these grasses will wither in the sun unless the earth, tilting on one of its axes, continues to pass storms over these lowlands.

Attended by these and unmentioned as well as unknown conditions, rice may grow among the grasses. The tidal areas of the

Asiatic continent include them, from the Yangtze to the Ganges and up the western shores of the Indian Peninsula. There the rice belt ends to the west with the droughts of West Pakistan. On the east, cool temperatures limit its northward continental extent to far up in southern Shantung and water-warmed Korea. Other species occur where the hot dry season has limited the forests to a scraggly growth or to scattered tall trees that can reach the deep-lying moisture. There rice species along with certain other grasses have adapted so as to complete their cycle of growth during the rains. Their dropped seeds lie dormant in the seared rubble until the drought is ended.

Some such sequences have extended into the past ten or fifteen thousand years, when gatherers trod these areas. Even baskets might be unnecessary, for some Negritos ate their crops as they ripened, with little attempt at storage. These gatherers of seeds could continue to live from their harvest, because they failed to pick an area clean. Thus varieties like *Oryza minuta* have survived, perhaps because the small seeds fall easily from the inflorescence, and elude the pickers. In Ceylon a perennial species demonstrated another way of surviving the ravages of the gatherers, for though its seeds were picked clean, it would grow again from the roots.

We can further specify certain social features of these gatherers. They must have organized themselves into small and rather mobile groups, if they were to feed themselves from the scattered and unpredictable stands of seeds. Because of their mobility, they could not handle large quantities. Thus, through many seasons, they depended on other sources of food, possibly hunting and fishing. In such circumstances rice was a seasonal addition to the annual diet, possible for the Hoabinhian and Bacsonian people in Vietnam, but probably denied because of climate to their cultural cousins in Yunnan and Szechuan. Whether they ate their rice as a welcome addition to their meals or as a sorry substitute for some missing delicacy, we shall never know.

Though the wild rices gathered in the years that followed remained an addition to the main sources of food, rice became a satisfying staple. The Munda once lived in the Ganges Delta region gathering rice varieties but subsequently took to cultivating the crop. In their present upland location, the hills of Chota

Nagpur, they are reported to continue supplementing their agriculture by gathering wild species. In southern India where the fine-seeded *Oryza officionnalis* grows, Burkill (1935) states, "the poor do not ignore it but tying the awns together before maturity save the grains for themselves (p. 1592)." Something of this sort seems to have occurred also in Thailand, if we rightly read the observation of Gervaise's account of his travels about 1680:

> Although it is only 12 or 15 years since maize was first sown in the Kingdom of Siam, nevertheless, great plains are seen covered with it. It grows so well on the higher land that we may hope it will soon be a fairly common product, but we realize that it will never be so common as rice which grows all over the kingdom so plentifully that the neighboring people send every year for supplies. There are three kinds of rice. One grows wild and does not require damp, marshy soil. The other two kinds must be sown by man (p. 6).

From this we presume that gathering of wild rice varieties again supplemented the cultivated crops.

How much of the wild varieties of rice these gatherers picked, how much they consciously left for the next year's seeding, we cannot say. The advantage of such gathering lay in the minimal demands of labor needed to acquire the food for eating. However, as the harvests only supplemented the ordinary diet of even a small population, we judge the size and dependability of this food source to be low.

SHIFTING CULTIVATION

When man abandons random gathering of rice, he must enlist nature's help to sustain himself. He must listen intently to the cadence of nature, learn the steps that correspond, and within his capacities perform the sequence that draws from his environment the greatest favor. Thus arose the clearing of fields by slashing and burning the forest cover, planting seeds at the bottom of a hole stabbed into the soft earth by a dibble stick, and, after a harvest or two, moving the performance to a new location. Here is rice grown like wheat or oats in dry fields watered only by the rains. The work must be timed like a dance to fit the rhythms set mainly by the composition and decomposition of soils, by the seasons, and by the cycles of vegetational growth.

Tropical soils have a slow underlying rhythm, for below the mountain outcroppings rocks must have weathered through stages of fineness to a granular size that can hold water, thus sustaining the roots and microorganisms essential to plant growth. Then plants begin to add organic matter to the sterile debris of limestone, gneiss, or granite. According to circumstances of exposure to wind and water, this soil can support vegetation for many years or few, but in the end the leaching of waters turns it into sand or laterite. Beaches and dunes testify to the sterility of sand. Laterite accumulates as a percipitate of aluminum and ferric oxides below the surface soil, where erosion of the overlying earth and contact with air turns it into a reddish rock, found in many parts of the tropics. Only in this period of transformation, between one form of sterility and another, can the earth sustain plant life.

Though the rhythms of soil transformation are slow enough to be translated by man into matters of place rather than of time, the stages, if not the sequence, are recognizable by shifting cultivators. "Rice," they say, "grows better in the dark soils, poppies in the red soils. The white and sandy kinds should be avoided." Here darkness refers to the good humus content needed for rice, and red indicates partly leached clayey soils, neutral or slightly alkaline, that favor poppies, while the whiteness or sand of leached soils is to be avoided.

The cycle of plant nutrients presents a wild beat in comparison with the rhythm of the soils, for it corresponds with the seasons. The leaf that rots only after a year or more of temperate-zone exposure disappears within days under tropical conditions. When rains are heavy and evaporation low, then the molds, fungi, and bacteria speedily convert the fallen leaves to humus. The soluble nitrates and phosphates quickly dissolve and are drawn deep through the porous soil. If a plant is ready at that moment, it will be nourished, but a week later the nourishing liquors may have sunk beyond the reach of its rootlets.

Having selected a site, the shifting cultivator fells the vegetation early in the dry season and allows it to dessicate in the mounting heat until, ideally, the debris is at a tinderlike state of dryness. Then fire is set roaring across the field. Properly done, this saves enormous work in clearing, destroys many seeds that might compete with the crop, and leaves behind in the ashes phosphates and potassium that help the growing plants. Cultivators must be sure to drill holes among the ashes and drop in a few seeds before the rains begin, yet not so far ahead that predatory creatures eat these newly-planted seeds. Should the burning fail to consume the forest debris because it is wet, poorly arranged for burning, or burned so soon that winds scatter the ashes, the crops will be in some measure dwarfed.

Once the seeds have sprouted, the new roots do not lie encompassed by humus as in temperate soils. Plant roots spread out through sterile layers in tropical soils, absorbing, as in hydroponics, the liquid nourishment washed down by the rains. Thus the best nourished plant in tropical soils extends its root system broadly and deeply to gather sustenance at various depths. Forest trees fit the cycle best, their fallen leaves and branches being quickly consumed by surface life and converted into nutrient fluids that percolate to the roots and reinvigorate the plants.

The shallow-rooted crops of cultivators break this balanced cycle. After harvest the spindly stubble and thin debris of the swidden fields and gardens cannot match the luxuriant leafy fall of the trees. Starved for nourishment, the microorganisms of the soil and the vital salts and sugars necessary for plant life dwindle with the thinning carpet of humus. A second year of planting brings reduced returns.

With new fields required each year, shifting cultivators need considerable amounts of land, heavily forested and lightly populated. Villages in the hill tracts of Southeast Asia enjoy these conditions to a greater or lesser degree, for every few years they follow the fields, moving their houses nearer them, and leaving in their wake abandoned clearings that gradually return to forest. Last year's field grows up with light-loving grasses plus *Ageratum, Eupatorium,* and other plants requiring little nourishment. Four years later bamboo has taken over, gradually shading the area so that the grasses and weeds disappear. If left undisturbed for twenty years or so, trees will take root in the shaded groves and grow to such heights that few bamboo stalks remain, and the soil is again replenished with heavy debris.

In lightly populated areas, villages of shifting cultivators may dance off in any direction, since they have little likelihood of crossing paths before the forest has returned. Only when other villages appear must they start to move within a confined area. Gradually, with greater population density, the pattern of their dance becomes circular, then ever narrower, until the village can no longer move. Then crops are rotated in fields outside a fixed village, and the dance is performed in stationary pirouettes, with only three to five years of fallow before reusing old plots.

Most of the shifting cultivators whom we encounter today have shaped their dances in narrowing circles. In his book, *Nous Avons Mangé la Forêt* (Paris, 1957), Condominas describes an old Mnong Gar of Vietnam returning to inspect a plot used 20 years previously. Vegetation had not grown sufficiently; they decided not to use it. A Meo showed me, on a trip through the hills, the fields that his village used 12 years earlier, but weeds rather than bamboo covered the hillside because another village had used them during the preceding year. The time was not yet ready for fresh cultivation. Elsewhere in northern Thailand, upland villages report returning to former fields after intervals of 3 to 10 years of fallow. They cleared away weeds and bamboo instead of timber. The forest had not returned, but enough nutrients had accumulated for one thin crop. Cultivators compensated for this hazard by clearing somewhat larger areas.

When the land is overtaxed because the period of fallow has been too short to regenerate needed fertility, nature responds with soil erosion and nearly ineradicable weeds. This latter signal of impending trouble is *cogon* grass (*Imperata cylindrica*) that rises six or eight feet over a tangled root mass resistant to hoe, mattock, and fire. In parts of the Philippines, Borneo, and continental Asia as well, this grass carpets a treeless and deeply gullied land. Many ecologists believe that the cycle of regeneration to forests has thus been permanently broken, and for evidence they point to hill tracts in southern China that have remained treeless for centuries. Here lies the terminal point for shifting cultivation, for though settlements may continue, economies must change to other techniques of cultivation or some other method of making a living, such as herding cattle.

Such a result ordinarily occurs when the human population has become too heavy. In the Philippines, according to Conklin

(1957, pp. 146–47), 130 persons per square mile sets an approximate limit. Freeman (1955, pp. 134–35), speaking of Borneo, offers a somewhat lower figure of 60 persons per square mile, while Gourou (1951, p. 31) estimates a tolerable density for the land of Indochina as only 32.5 persons per square mile. In 1960 only 13 of the United States exceeded a density of 100 persons per square mile.

While the figures for Asia doubtless indicate points at which danger may be expected, they fail to indicate variations, for the rate of soil regeneration fluctuates with the soil quality, the size of area cleared, the moisture, and the number of days of sunlight. A small field cleared in the forest returns more quickly than a patch of identical size surrounded by bamboo and occasional standing trees. We know villages of shifting cultivators in northern Thailand that have stood at a single site for 40 years, where people return to old fields after only 5 or 6 years of fallow. On the China–Burma border, fortress villages have stood for centuries at critical mountain passes exacting tolls from passing caravans, the people supporting themselves on fields that could not have lain fallow for more than 10 years. Elsewhere, because of granular soil texture, erosion rarely occurs, despite raw gashes and repeated use of steep hillsides, and in other areas *cogon* grass seems not to take hold, perhaps because local species of weeds can compete successfully with it.

The fact of shifting cultivation per se tells little about how its cultivators live. As a technique of agriculture it often supplements a considerably broader food quest. In the rice fields themselves, these people plant scatterings of maize, gourds, squash, beans, peppers, yams, and taro, to begin an inventory that never quite ends. There are "partial" shifting cultivators whose main mode of support is fishing or even raising irrigated rice. The famed terraced mountains of the Kalingas and Ifugaos in the northern Philippines are interspersed with patches of rice in swidden fields, for irrigation water does not reach every corner of the mountainside. Many coastal villages in the Celebes or on some craggy island off Sumatra turns to the sea during the dry season but plants rice on the hillside during the rainy season. Many of the "integral" shifting cultivators, for whose diet rice is a fundamental ingredient, hunt for birds or deer in the surrounding forest, and possibly grow opium for sale. The partial shifting cultivators tend to live in permanent villages, but not all the integral cultivators live in seminomadic fashion. The Lahu Nyi uplanders of Thailand, Burma, and Laos move their houses every few years to be nearer their fields. Their unconcern for permanency shows in a simple style of house-building, the absence of plan in village settlement, and vague claims of territory in the surroundings. In contrast, the Akha, who are also "integral" shifting cultivators and occupy the same area, erect elaborate houses, allow in the village plans for a street and dancing area, specify areas in the surroundings for cultivation, and sometimes reserve other lands for forest. They construct their villages with the intent of remaining settled for some years on a given spot.

Shifting cultivation can be found within the rice-growing area of Southeast Asia from the Himalayan ridges at 7,000 feet down to the extensive sea coasts. This is the same general technique that supports yam and taro cultivation among Melanesians and Polynesians, maize in the Americas, and millet in Africa. Here is a worldwide technique good for growing what grows in the area.

BROADCASTING

Where the dry rice of shifting cultivation depends on the decomposition of organic matter deposited on the soil, fixed fields have usually lost the supply of vegetable debris. Plants would starve were the essentials not conveyed to them. Thus rice in settled fields usually depends on water to bring its nutrients, much as the blood stream carries sustenance to the deepest-lying cells of the body. Organic matter as well as minerals come when streams pass forests and villages on their way to the rice fields. Though some organic matter also accumulates from past growing seasons in the clayey soil, essentially water becomes a life-bearing brew, the *soma* of rice. A flooded field is short on oxygen as compared with a well-cultivated barley patch, yet oxygen becomes available through the bacteria that break up the organic products of fermentation. In this dank airlessness nitrogen is converted by other bacteria into ammonia rather than more familiar nitrates. In the presence of ammonia the brew tests more nearly neutral or even alkaline than acid, so that phosphorus becomes available to plants as ferrous and manganese phosphates rather than the more familiar phosphorous acid. Unlike roots in dry soil, the rice roots at the bottom of a flooded field serve mainly to anchor the plant, while the higher rootlets drink in the necessities of growth.

The watery element, the *Yang* of old China, states the limits of fixed field cultivation. Here is no longer a dance with frequent improvisations, but the timed moves of a courtly etiquette, a Confucian bow in response to a bow, a nodded smile in response to an enquiry. The enduring barometric lows moving from the Asiatic mainland southward beyond the equatorial islands signal the ceremonial beginnings. So winds that set kites flying move in April from the southwest and in December from the northwest, while cultivators seek to break the drought with water-oriented rites. As the heat mounts, uneasy winds mount, crashing waves against breakwaters and casting loose palm fronds along the shores. Clouds tower on the horizon, piling higher and higher, so that in a few days or a few weeks the first rains fall, changing the cycle from evaporation to precipitation.

Then drought is crushed, and the life of the new year emerges. Withered browns turn to green; frogs, fish, and toads, which had survived in the deeper layers of earth, burst forth. Little wonder that poets of the region have loved these moments; an ecstatic Burman wrote (Burma Research Society, 1959):

> Celestial drums send forth a mighty roll and the fairy conch swells the pealing note. The luster of the lusty sun is quenched in wreaths of cloud and the jocund sound rings through the azured vault like the thundrous rattle of arms. The god of the sky sends sheets of swift rain. Pools and lakes and woods, marshes and streams and springs brim over from countless rivulets. Rivers and streams look ravishing in a thousand charms. . . . (p. 9).

Cultivators have learned to respond to these signs in the sky and earth, so that the offering of seed on the soil will be blessed with maximum growth and be ready for harvest on the day when the last storm of the year has passed. They have learned the etiquette not only of soil and seed but particularly of water.

Broadcasting, the scattering of seeds on the surface of a field, contrasts with two other techniques of planting. We have already seen shifting cultivators drilling holes in which to drop a few seeds, and in the following section we shall describe transplanting from a seedbed. Yet the term "broadcasting," or *naa waan* in Thai, *srauv prous* in Cambodian, *padi taboran* in Malay, implies more than a mode of sowing seeds. It refers to a complex of agricultural techniques associated with particular terrains and conditions of watering the crop. Fields are open and undyked, so that water comes by natural flooding rather than by irrigation. Plowing and seeding are apt to take place while the field is still dry, before the rains have fallen. When properly launched toward maturity, the plants often appear irregularly spaced and studded with weeds.

This mode of growing rice fits well certain terrains. In a broad valley the rains may come to any particular area in due course, but heavy rains upstream may swell the stream level days and weeks before local showers have occurred. So the nearby river rises and overflows its banks, providing ideal conditions for rice cultivators. But how high will the water rise? If too high, the plants will be drowned. How quickly does the flood come? If slowly, the plants will be tall enough to survive a considerable depth of flooding. How long will the water remain? If it suddenly ebbs away in the middle of the growing season, the crop fails.

Cultivators must have the seeds ready to grow when the first drop of moisture reaches them. As the fullness of the grains depends on the duration of the growing season, each householder seeks to stretch the growth to its ultimate. So while the sun still beats down on the cracked earth, the plowman begins driving his buffalo or oxen to pulverize the crust, then breaks the clods with his harrow. When seeds are scattered and lightly covered with dust, he may await the rains.

The plowman must, however, have heard the voice of nature well before this final act, so as to read the slope of the land and its flooding and then select the proper variety of seed. Where the flood remains for perhaps only a week or two, there stands the high point of land. Back from this point toward the deep water in the center of the valley, the flood remains for three or five months. Where the water reaches a depth of 5 to 10 feet and is last to dry, he must recognize the conditions for planting "floating" rice. If this kind of seed can rest but a month in the mud to begin its growth, then the crest may rise several inches each day without danger of drowning the crop. As these varieties mature slowly, they are the first to plant and the last to reap. Where the duration of the flood is less, other varieties fit because of their tolerance for shallower water and their shorter period of maturing. Like courtiers choosing the proper spot and hour for the king to pass, and waiting for the proper moment to state their requests, these cultivators time the spot and hour of planting.

The flood waters rise early or late in accordance with the currents of wind in the seasonal oscillation of storms, with the confluence of streams, with the slope of the terrain and the permeability of the soil. The great flood plains of the Arakan coast in Burma and the expanses around the Tonle Sap in Cambodia exemplify such broad shallow bowls for broadcasting. To this may be added large sections of central Thailand where the Chao Phraya River overflows its banks. Yet broadcasting may, according to terrain, mix with other modes of cultivation. Moerman (1968) describes a village in Chiengrai province of northern Thailand where the river's edge is given over to broadcasting while the higher land contains irrigated fields. Nor may we forget the grander cycles lying behind these auspicious circumstances: the weathering of mountains into clay, the forests on the higher slopes that enrich the crop with their water-born debris, the changing axis of the earth now sending storms over the Far East but someday sending them elsewhere, and the uplifting of the mountains that have located the valleys but eventually will bury them.

TRANSPLANTING

In transplanting, a single feature of an agricultural complex has been singled out to indicate the whole. The Thai word *naa dam,* the Cambodian *srauv stoung,* and the Malay *padi chedongan* likewise single out the moving of young rice shoots from a nursery and setting them out in a larger field to grow and produce the grain. Essential to this technique is control of the water, and here the technique depends on whether the water source is rainfall or an irrigation system.

The first alternative to natural flooding is the simple dyking of a field into a basin where dependable rains can be held for the period of growth. Thereby intermittent but necessarily ample rains can be guided to sustain the 90- or 120-day rice varieties without damage of flood or dryness. The height of the dyke sets the limits of the water's depth; if the water rises too quickly, a few strokes of a mattock can quickly lower it to safe limits. Its particular disadvantage lies in the relative sterility of its rice-growing brew, if unenriched by the accumulations of rivers in passage. The salts, mold, fungi, bacteria, algae, and organic debris must then come from the air or the soil of the field itself. We learn further that this brew sours when rains are not sufficient to dilute and drain away injurious by-products. On these accounts the productivity of such fields is low. In Thailand rainfall fields occur where soil may be good but lies beyond the reach of irrigation. Ordinarily they are less desirable sites, but in Burma we learn that much of the bounteous rice crop of the Irrawaddy plains is grown in this manner:

> One rice crop only is grown each year. It is planted out from seedbeds in July, soon after the early monsoon rain has moistened the soil enough for ploughing; rain water accumulates in the fields as the rice grows. No river water is used; by December the mature rice is standing in dry earth ready for harvest (Rawson, 1963, p. 220).

With straw stubble burned in the field, sufficient water and probably also green manuring to enrich the growing rice, a considerable crop lies within the grasp of a discerning cultivator.

A continuous source of flowing water, which can be sluiced into any dyked field and then shut off, offers greater control and is, of course, preferable. Taken from a river or mountain stream, the water holds the silts and organic debris that feed the bacteria and algae necessary for plant growth. In the Sarapi district of northern Thailand, where the Mae Ping flows year-round into irrigation channels, a wooden plank directs the flow into any field. More often the waters flow too high or too low to make such an easy junction. The Red River of North Vietnam winds like a dragon through the rice fields, threatening to engulf, should its levees break during the flood period. Though Thailand's Chao Phraya River valley is pronounced excellent in contour for uniform flooding, even there rains may come late, and farmers have to flood their fields by sending children to push the treadles that splash water up a trough into the fields. The variety of these devices is awe-inspiring, ranging from the simple water shovel suspended in a frame to the stately *noria* where the river current lifts each water-filled bamboo tube on the paddle wheel and empties it hour after hour into an irrigation canal.

Transplanting is a ritual demanding not only the precision of a dancer and the responsiveness of a courtier, but the orderly liturgical understanding of a priest. After learning the watering of fields, cultivators become ready to respond to other subtleties of the environment. Seeds as always must be fresh. Those collected at harvest must be returned to the soil before the germ dies within the grain. By selecting only the longest, roundest, or largest ones to continue the cycle of growth, more bounteous crops are earned. Out of selection have come seeds with special attributes such as those that respond to the shortening hours of daylight to produce their inflorescence. These harvests follow the equinoctial passage of the sun over the Equator. Others of longer or shorter periods mature and bloom like fetal animals in the womb, heedless of the heavens.

Cultivators learn the subtleties of soil and plowing. Soil exposed to sun and air after long inundation changes from an easily pulverizable substance to something almost as resistant as ceramic. After dyking, plowing is impossible until rains soften the soil, though formerly they may have plowed in dry earth. They have learned to plow only a few inches deep where organic matter lies thickest, and none of them explain this practice in terms of neutralizing the sterile alkaline layers of soil below with the acid soil of the surface. Yet all cultivators know that the cloven hooves of their draft animals and the sharp-nosed plow help the soil to breathe again. Their metaphor contains the responsiveness of nature, and cultivators have observed nature's heartier response when the seeds are sprouted with special care. They select a special field as a nursery, lying near a ready source of water, pulverize the lightly flooded soil free of clods, and stir in fertilizers. Some practice a further refinement by germinating the seeds in a dark moist spot before spreading them on the nursery bed, so that only seeds of demonstrated power of germination are allowed in the tightly packed nursery. A few weeks later, when the gangly seedlings are pulled from the nursery, trimmed and set to grow in the large field, spacing and transplantings become important. The seedlings must lie far enough apart to prevent competition among the roots that might starve a neighbor, and near enough to prevent each other from falling into the water when strong winds blow. When every corner of the field bristles with evenly set plants, nature responds more kindly to the offering.

Even with careful weeding the nurture has not ended, for pests and diseases must not interfere. The leaf caterpillars and beetles, the stem-snipping crabs and other visible despoilers must be promptly countered. The invisible ones like the *Brusone* fungus that rots stems, or the worm borers hidden within the stem, must be dislodged before the damage spreads. The birds that come to eat the ripening crop are driven off with noisemakers and swirling twirlers. The rats that come in droves to devour the unharvested grain or an elephant that tramples down a whole field bespeaks refusal of a petition and counsels patience.

When at last the yellow sheaves are carried to the thrashing floor and the new harvest is being stored in bins, appreciation is variously expressed. In ritual discourse, we cannot deftly specify whether the gods and spirits are one or many. We hear a medley of names—the Rice Mother and the Lord of the Earth; the rising river and the nitrogen rich fertilizer; Indra, Allah, and the Lord Buddha—and there are many more. The Thai cultivators of wet rice address Mae Phosop, the maternal Rice Goddess:

> O, Rice Goddess, come up into the rice bin. Do not go astray in the meadows and fields, for mice to bite you and birds to take you in their beaks. Go to the happy place to rear your children and grandchildren in prosperity. Come!

MULTIPLE CROPPING

The term "multiple cropping" refers to a sequence of crops grown on the same plot. Reaching a particular stage in the growth of a first crop, perhaps harvest time, a week before or after, becomes the signal for planting a second crop, which, on reaching a point in its maturation, signals the planting of a third, and so forth. The timing may be leisurely enough, as with the transplanters of northern Thailand who devote the two or three months after the rice crop to growing tobacco and garlic with a month or six weeks to spare between crops, time well used for cutting wood or repairing irrigation ditches. The cultivators in Chinese Yunnan, however, must hurry with their rice harvest in order to plant winter barley and then must hastily harvest the barley in order to plant the spring rice. Rice growers of Chiengmai province, Thailand, with their two crops per year produce more per acre than in any other part of Thailand, though in Taiwan and Japan many spots raise three crops of fast maturing rice.

Not every valley is suitable for agriculture of this intensity. Water must be available throughout certain seasons, as it is in the irrigation systems of Japan and Taiwan, but in only limited parts of Southeast Asia. In Cambodia the great lake called Tonle Sap forms a natural reservoir for holding a portion of the summer floods brought down the Mekong River. When the crest has been reached in early autumn, gates are closed, and the water is saved to drain into the fields through the dry season for a second crop. In Burma much the same principle was used, when reservoirs fed by the rains were built. The *noria,* invented in China, can fill irrigation ditches of limited size as long as a river is running fast enough to turn the wheel and pour the water into a reservoir. So far electric or motor-driven pumps have had limited use, for the engineering problems involve storage and distribution no less than power to raise water in this alternately wet and dry region.

Far from what one might expect, multiple cropping, even with abundant water, is not a mechanical repetition of planting and harvesting a first crop. The light for the summer crop varies from that of the winter crop so that photosensitive rice cannot be used in both seasons. Temperature sensitivities also differ, so that varieties depending on warmth for certain kinds of development cannot be used. With continuous or near-continuous use of fields, the fertility that comes solely from annual inundations no longer suffices. The field workers who gird themselves once a year for an exhausting effort tolerate working conditions differing from those who perform the same drudgery two or three times per year.

To develop the requisite new techniques without benefit of experimental stations or extension workers calls for more than precision of timing, routine judgment, and responsiveness to ordinary needs of the crops. Besides observing the plants for signs of yellowing or of retarded development, a good observer notices the amount of scum growing in the water, perhaps a strange sort of spider building a web between the plants, or that the great white egrets have moved this year to wade in a new area. Such kinds of awareness may have been half-expressed in the now ridiculed signs that guided older generations to plant in the light of the moon or after the swallows had arrived. Beyond this lay other guides harder to articulate: a certain degree of grittiness in crumbling a clod of earth, a certain pungence in the fragrance of a crushed leaf.

However the means for double cropping were discovered, we may observe some of the methods that are used. In Vietnam's Red River Valley the rice crop is neatly dovetailed with garden greens and squash grown on the dykes, while in the dry season drought-resistant wheat, seeded during the diminishing rains of November, occupies fields too high (and thus too dry) for planting a second crop of rice. Japanese cultivators learned to plant grains and tubers at the proper moment in the fields so that the rice could be harvested without injuring the growing second crop. As for fertilizers, villages situated in areas with unused hillsides have developed animal husbandry, thus enabling them to spread their fields with animal dung as well as with village night soil. Periodic rotation of crops with nitrogen-fixing soybeans have proven successful elsewhere, and in the Red River Delta, selected villages have abandoned second crops of rice for an alfalfalike green that can be sold as green manure to be plowed into the soil. Chemical fertilizers, always limited by their expense, have been tried and sometimes found successful. They presume good control of water flow and level in the field so that fertilizers do not wash away before becoming effective. Even then certain rice varieties are unresponsive; the presence of extra potash and nitrogen seems not to stimulate their growth. The problems of labor have been worked out in Java perhaps by pure necessity in this crowded land; seeding, transplanting, and harvesting go on stimultaneously at all seasons.

In mainland Southeast Asia these multiple cropping systems with few exceptions occur in only the most heavily populated regions, such as the Red River Delta. The homelands for these techniques are certainly China and Japan. Bang Chan has never tasted this degree of complexity and responsibility.

HIGH-YIELDING RICE

As developed in nineteenth-century Europe, natural science introduced a vision of the world as the varying shapes of transformed energy. Magnetism, electricity, heat, light, and matter represented but differing forms of energy, each holding the potential of being converted from one to the other. So too plants may be seen to grow by transforming light and heat from the sun into vegetable tissue through photosynthesis.

During the 1950s, faced with a likely worldwide famine if population increased faster than food resources, many agricultural scientists turned their efforts to increasing the food supply. Among them, Japanese rice experts, setting up shop with Colombo Plan support in Calcutta, crossed many varieties of rice in search of a special kind. In order that it might grow at any season of the year, it should be nonsensitive to photo period. It would respond to fertilizer and reach maturity quickly. Yet most important for energy transformation, the plant would be short-stemmed; thus it would not expend energy on growing a stem, and it would allow the growth nutrients to increase the numbers of grains on each inflorescence. Among hundreds of hybrids, approximations of this ideal, which performed miraculously when subjected to special growing conditions, were found. As if the rice were an Alpine plant having to complete its cycle to maturity before being covered by the first snows, the new varieties were planted late in the seasonal cycle and grew in 90 to 100 days, rather than in the 150 or more days required by many varieties. Fields could therefore be made available for other crops, if not for rice. As for yields, they more than doubled, often tripled and quadrupled, depending on the standard used: 2.9 to 5 short tons per acre vastly exceed the more traditional 1.0 to 1.5 short tons per acre.

While these researches focused on *Indica* varieties, the International Rice Research Institute developed parallels for the *Japonica* varieties. At this station in the Philippines, researchers speak of growing rice during the dry season when more solar energy pours down upon the earth from clear skies. By collecting water in reservoirs during the rainy season and planting under the full rays of the sun, a cultivator may greatly improve his yields.

These remarkable hybrids, products of the concept of energy transformation, have demonstrated the possibility of meeting the world's food shortages by increased grain production. To achieve this possibility, however, presupposes reorganization of the habits of millions of people. Few now control water with the required precision, and many cultivators cannot afford to buy adequate quantities of fertilizer. More boats, tools, space, and people are needed to handle the enormous crops, and the bulk that once could move smoothly along canals may require additional railroads and highways. Millers accustomed to particular pressures to burst the husks will fracture the new grain until they learn to mill it properly. And people must accustom themselves to its peculiar aroma and taste, which is not immediately agreeable to practiced rice eaters.

POSTSCRIPT

As described, these ecosystems may sound like models, though in fact they are mere paradigms. The variants flow thick, as one begins to observe the local vagaries of water, soil, climate, seeds, and population. Moerman (1968, p. 53) describes dwellers of northern Thailand drilling holes rather than broadcasting seeds in the nursery before transplanting into irrigated fields. These people may be taking advantage of greater organic richness in the thinly forested area where nurseries are located. In Cambodia rice plants are set out in pools of water that remain through the dry season. In northeastern Thailand, where water gathers into pools only during the rainy season, these become the rice fields for the villages. Louisiana rice is drilled by machine into prepared fields, the same means used by the shifting cultivators, yet these fields are dyked for flooding. Californians broadcast their rice by airplane into fields where water level is controlled by pumps and spillways. In Japan, ever anxious for larger harvests, the rice seeds are broadcast between furrows of winter wheat and barley; these latter crops once harvested, the field will be flooded. In Israel, rice seeded by drilling is watered not by flooding but, like any garden crop, by spraying. So man listens to the variations in nature's demands and carefully chooses his reply. In Bang Chan, we can only trace the changing ecosystems as far as multiple cropping, which in the 1950s still remained embryonic.

Chapter 4

THE ENERGY REQUIREMENTS OF SOCIETY FOR RICE CULTIVATION

Having indicated the ecosystems that bring rice and man together, let us now observe an important ecological difference between gathering rice and cultivating it. Gatherers expend no energy in producing rice. As long as they, by accident or design, leave seeds enough to perpetuate the plundered rice patches, natural forces independent of man continue to promote the growth of another stand next year at the same place. Whether man, deer, or bird eats the product makes little difference to the plant community. We describe such a relationship between these creatures and the rice they eat as a portion of their various ecological "niches."

The ecosystems of rice that involve agriculture differ from gathering in that they require sizable expenditures of energy by a cultivator. His work alters many of the local plant and animal communities together with the natural cycles on which these communities depend. Such an alteration may be temporary, as when shifting cultivators plant for a year or two but then, by abandoning their fields, permit the natural communities to return. The other kinds of ecosystems of rice presume a more or less permanent alteration of many natural cycles, and if we consider weeds in a garden or rice field as the first step in a succession returning toward some variation of the preexisting natural communities, then more than half of a cultivator's work goes toward maintaining the environment necessary to grow a crop.

The actual planting and harvesting of rice takes relatively little time, but weeks must be spent preparing the soil and weeding in order to block the natural succession that would return the area step by step to savannah or forest. We shall call the total of these active working relationships needed to produce and maintain an environment for man as agriculturalist a "holding." Without the plowing, harrowing, and weeding, without maintaining the walls of dyked fields for transplanted rice together with the irrigation ditches, without the shelter that the cultivator builds and the societal arrangements that bind him to his neighbor, the crop would be impossible.

A "holding" differs from a "niche" in that it presumes active work by man to transform, maintain and thus to utilize the natural environment. In ordinary parlance we use the word "holding" to mean something grasped as well as the act of grasping; a third sense relates it to property, especially in land. Through its Anglo-Saxon roots we trace a connection to the Germanic word *Held,* meaning "hero," with its connotation of defender of a domain, from which our schoolyard game "King of the Hill" derives its meaning.

A holding presumes a unique kind of relationship to the environment that is not specified by other terms in ecological vocabulary. Though man competes with the weeds to maintain a cultivated area and consumes the crop like any other predator, man's holding alters the environment for the plants he raises as well as for himself. In this same vein the relationship of man to domestic plants and animals is symbiotic, but in an active manner that contrasts with our passive symbiosis to mice and cockroaches. From these considerations, relationships in a holding exceed the limitations of the word "niche," which a plant or animal transforms at its own peril, as parasites may kill their host.

If definitions help understanding, it may be useful to say that a holding is the totality of relationships of a given species or community of the species to an environment partly or wholly altered by that species for the purpose of survival. Of course, man is not the only creature that transforms its environment. Coral builds lagoons and reefs that harbor fish unable to cope with the open sea, but this effect is a by-product rather than a necessary condition for the polyps to survive. The ants, termites, and bees that build enormous mounds as well as smaller hives may be said to have staked out holdings because their survival depends on these artifices, and in this sense the nests of birds, squirrels, monkeys, and apes are also transformations of their environment, as well as the burrows of hedgehogs and prairie dogs. Somewhat more enduring and architecturally impressive is the dam built to sustain a colony of beavers. Thus man is not the only creature that survives by transforming his environment, though he is one of the more active at this work.

Finally let us note that each holding "rests" upon a niche. Though transplanters lead water through irrigation systems to their fields, they ultimately depend on the rains and the immutable changes of season to bring water into their canals. At this point they must adapt themselves to the fixed conditions of nature, much as the gatherers plucked wild rice where they could find it. Similarly, when man mines for metal, quarries rocks for houses, cuts natural stands of trees for timber, converts wind power to a form that turns a mill, he meets the fixed dimensions of nature and must adapt to them. In this sense each holding depends on forming a variety of complexly mediated relationships to the environment and thus presupposes an underlying niche. However, since man's efforts are concerned with both the transformed and the untransformed aspects of environment, and since man, in cultivating rice, deals primarily with the transformed, we are constrained to consider primarily holdings rather than niches.

We ask what is demanded for a group of people to produce rice in one of these holdings. To gain an answer we must seek to determine not just that a given group practices shifting cultivation or broadcasting, but also the group's degree of dependence on its crops, its resources of tools and workers, and in particular the social requirements of cultivating in a particular manner.

Social groups are only in part reservoirs of workers. A moment's consideration shows that the work force of any society is limited by various social and cultural as well as environmental factors. Even the slaves, who built the pyramids under persons whom we imagine to have been the harshest taskmasters, could not be expected to work indefinitely. Storms and darkness brought the slaves respite from their labors. No superintendent of construction could threaten the very completion of his project by failing to provide food, drink, and rest, even if he limited the time to make merry. Nor could he risk slowdowns and revolts among people who had to move tons of earth by failing to provide at least the elementary tools of the day. Social norms of Nubians and Canaanites, beyond the ken of an Egyptian master, set standards of work and enforced them with sanctions. These limitations apply to almost any group, and for an autonomous one there can be added minimally the limitations due to age and sex, status, and specialty of the workers.

To lend a certain order, we have chosen to couch our observations in the metaphor of energy exchange between a given social group and its environmental holding backed up by its niche. Here the particular holding, rather than human society, becomes our reference when we refer to its input requirements, that is, what a social group must do in order to gain certain outputs of rice. The energy metaphor further gives us the advantage of comparing quantitatively the inputs and outputs of various holdings on the basis of man-days of work and the resulting tons of provender. All who have experienced other cultures know that five man-days of work by Burmans are not necessarily the same as five man-days of work by, shall we say, Peruvians. Similarly, a ton of grain may not support a Peruvian as long as a Burmese family. From one culture to another the units are not exactly comparable, and this is also the case with various modes of cultivation together with their varying ecological holdings. The prospect for scientific precision is disheartening, yet in the absence of better standards we shall continue on our way, assuming it better to have scanned the scene with distorted lenses than never to have looked at all.

In the first section below we shall consider the organization of societal input into shifting cultivation, broadcasting, and transplanting. Multiple cropping and high-yielding rice, not having come to Bang Chan, are at most incidental concerns. A second section will compare the pattern and total input required by each of these modes of cultivation with the output. This is followed by a summary of Chapters 3 and 4.

SOCIETAL ENERGY AS A QUANTUM

Any lowland rice grower and many uplanders can tell how much rice they must have to feed their households through the year between harvests. The figure, varying somewhat for locality, comes close to 1.1 lbs. of unhusked rice per day or 400 lbs. per person per year. With children eating somewhat less than adults, even setting aside as much as 110 to 132 lbs. for seed, the average household of about five persons is well supplied when it has 1,980 lbs. of unhusked rice in its bins at the beginning of the year.

We may take this figure casually, remembering the varying meanings for each group. Uplanders with a game and root-filled forest standing above them on the hillside are less dependent on their crop than are lowland dwellers on the plains, where fish and an occasional snake are the main dietary alternatives. Add to these the people who are wasteful of their resources and those who are frugal, or those who would conceal their harvest wealth and those who proudly announce it to the world.

Whatever cultural and environmental qualifications shade the measurement of crop, the size of the harvest is the important measure. One household may use twice as much land as the next, perhaps twice as much seed, and work twice as long, but as long as the harvest fills the bin adequately, no one complains. While researchers calculate and compare yields, growers find these figures meaningless. The days worked per unit of land or per unit of crop have little relevance to the practitioner, who must do his best with the land, tools, hands, and weather that have come his way. Not even the most seasoned and successful producer of market rice can anticipate the price for his future crop, calculate his annual income, determine the rate of return on investment, and emerge with a balance sheet of assets and liabilities. He knows how much he owes on credit, and when the money for the crop is in hand, he pays off his creditors as quickly as possible and spends the rest for clothing his children, buying a boat, or cremating his mother deceased these many months. There are no savings buried under the floor, for keeping money in the house invites robbery. This is not the language of the rice fields.

Ask a grower, however, how much land is necessary to raise a crop of a given size, how much seed must be sown, how many people are required for each step of cultivation, and what tools. These features he knows, and well he may, for they represent the wisdom of personal experience reinforced by the social standards of his group.

LAND. An upland cultivator of shifting fields can say that he looks for land of black or brown color, with a fair portion of clay and as little sand as possible. By the luxuriousness of the vegetation he knows whether the soil is fertile; by the kind of vegetation in the succession of regrowth, how many years ago the last crop was grown. With the required size of the harvest in mind, he judges how large an area to clear. Doubtless he has already set aside the amount of seed that will be required. Beyond these calculable indicators lies a host of tenuous judgments: the slope of the land may be steep or gentle; the surrounding hills and trees may shade or leave the clearing exposed to sun and prevailing winds; predators and competing plants may live in the vicinity. Because weighty judgments on these matters require at least a year of observation, one can understand why rice growers hesitate to move to radically new spots and why even shorter moves may be hazardous.

Fixed cultivators of the lowland, working where land is held in small units, have little choice of place to grow their crop but often can determine how much they will cultivate. An owner with more land than he can use may rent out a portion to a neighbor with too little. Each seeks to harvest at least enough to feed his own household and as much more for the market as possible. How large a crop this is depends on the household labor available. As children grow toward maturity, can carry adult-sized loads, and paddle boats with adultlike precision, the labor force increases, and a parent rents a little more land. When the children leave home to marry or take their vows to enter a monastery, the household cultivates a smaller area.

LABOR. Shifting cultivators need labor beyond the household only at one time of the year. To fell the trees and clear the brush from a forest requires the strongest and ablest men. The Meo and Lahu tribesmen of northern Thailand, having previously agreed on the spot to be cleared, summon all adult males to chop and hack. On another day a few months later they jointly set fire to the tangle of dried limbs and branches, watching the flames roar up the hillside. Thereafter the individual household clears up the unburned residue and carries on by itself for the remainder of the season.

By applying certain guidelines, the individual household that broadcasts seeds on its fields can also carry on much of the time without extra help. A *rai*, the unit of land area in Thailand, Laos, Cambodia, and the Shan States of Burma (equaling 0.4 acres), is the area that the average buffalo can plow in one day. This rate guides plowing so that ideally the land will be ready for seeding at the latest on the day before the first rains. Only at harvest time do demands for labor exceed the household supply. Then the grain must be cut, packed into sheaves, and carried to the threshing floor while the kernels still cling tightly to the inflorescence. If there is delay, the few days of dessication turn the connections brittle, so that grains drop in handling and are wasted. To finish in one day the job of cutting and moving the grain from the field to the threshing floor, one applies the formula: 1 man harvests ½ *rai* per day. Where neighborhoods exchange labor, returning one day of work for each day a neighbor spends in one's own field, a cultivator can count his days of labor by the number of persons he must invite to harvest his own field.

Transplanting requires extra workers at two other periods as well. The day the nursery bed has been planted with sprouted seeds fixes the date for transplanting 30 to 40 days hence, according to the rice variety. By then a man must have plowed and harrowed all the fields into which he will transplant the seedlings. If the early rains have already come, the owner can, with his one team of buffalo, prepare as much as 20 *rai* (8 acres), for draft animals must be allowed a day of rest after six or seven days of work. The field is harrowed perhaps at the rate of 2 *rai* per day. If the rains needed to soften the packed earth of the field before plowing chance to come a week or two after the seedbed has been planted, the cultivator must telescope the work into shorter periods. Most growers in Bang Chan stagger the plantings by four or five days and use differing varieties in order to extend the time for plowing. An extra worker may be found for the occasion, but unless an extra team of buffalo is also on hand, he will be of little help. But as every animal across the plains is hard at work, there are none to borrow.

Uprooting the seedling from the nursery bed for transplanting is another period when work becomes compressed into rigid time limits. The plants will grow no more in the nursery and must be torn from the mud, bundled for carrying, and, after clipping off a few inches of green shoot, replanted in the broad field, all within three days if the seedlings are to survive. Workers gather 200 bundles in a day; once the bundles are taken by boat or carrying pole to the fields, the newly flooded fields are planted, one-half *rai* per day. Finally, like the broadcasters, these transplanters too must calculate work requirements for harvest.

In summary, each mode of cultivation has its particular pattern of peak labor demands: shifting cultivation, a single period at clearing the forest; broadcasting, a single but considerably tighter demand at harvest time; and transplanting, three periods of high labor requirements during uprooting, transplanting, and harvesting, with possibly a fourth during the plowing season.

CAPITAL. Each mode of cultivation must also have its capital supplies of land (if purchased or rented), working facilities, and tools, and every society has its set procedures for acquiring them and its own rates of manufacture. As long as there are underpopulated and uncontested corners of the realm, land may be acquired by shifting cultivators who simply squat on some patch and hold it as long as they continue to work it. For their work they must have an ax blade, an iron tip for a digging stick, a mattock, a hoe, a brush knife of the machete type, and a small finger knife or sickle for harvest. A man can fashion his own handles from wood in the forest, but blacksmiths in the locality make the tips and blades from steel purchased by the customer from the lowlands. Fifty cents or a dollar will buy enough steel for most all of these implements, and in a day a blacksmith can fashion about a dozen, depending on the initial shape of the metal. His services are usually paid for by a meal with meat, a day's work by the customer pumping the bellows at the forge, or both.

A man with land for broadcasting must have many of the foregoing tools plus plow, yoke, harrow, buffaloes, and a cart or boat. Nor may we overlook the pitchfork, brooms, and winnowing equipment for harvesting and cleaning large volumes of grain. Should he turn to transplanting on the Chao Phraya plains, he would add at least a dragon-bone pump, driven by a windmill or possibly a gasoline motor, and some sort of box or old boat for sprouting seed rice before sowing it in the nursery. Though he may make his own plow, yoke, brooms, rakes, and even boats or carts, most items are secured in the market.

Without these tools the grower is helpless to carry on, as debtors forfeiting their equipment and livestock to some creditor know to their sorrow. At the moment of loss they change from growers of rice to laborers. If the market price of labor in manufacture were translated into days of work, they too represent definable amounts of human energy necessary before a crop of rice can be grown.

From these observations on societal input of labor and capital, we find not a limitless reservoir of human energy but an organization with its own limits and own rate of energy expenditure. One can, of course, work a water buffalo, one's children or one's self around the clock for short occasions, just as gray beards in a pinch can run a gasping mile. For the constant haul such exertion cannot be sustained. So a society develops its own standards of work, its quanta of energy input which tell a man when he should feel tired, which distinguish adults from children and the aged, and which define both work and rest. According to the rigors of the setting, some societies make greater, others lesser demands upon their members. Boating on the placid canals of Bang Chan would seem like child's play to Phoenician and Viking sailors, yet these people too defined their standards for a voyage in days for rowing and rest, in times to eat and drink. Over decades and generations these rhythms may drift or be driven toward slower or faster beats, but every good boatswain knows the limits of his crew.

With these many standards of group work in mind, it seems possible to say that *a society's quantum of energy patterned for greater and lesser inputs through the seasonal round must match within tolerable limits the requirements of the ecological holding.* To this we add the corollary that failure to maintain this patterned quantum of energy input, by either excess or insufficiency, brings about a decline in the output. Applying these propositions to rice cultivation, we seek to say that each group of growers must become adjusted to expending its energy in the proper amount at the proper moment, as the case may be—to clear the forest, plow the field, or harvest the crop. To vary significantly from this optimum amount leads to a diminishing output of rice.

Geddes (1954) relates the following from his study of the Land Dyaks of Sarawak: A, with 4.78 acres of land and 224 man-days of work, produced 268 gallons of rice. B, with 1.8 acres and 184 man-days of work, produced 480 gallons of rise. He explains as follows:

A, the owner of the field, began the season in style by engaging helpers for a total of 75 man-days to clear an area. Had he been able to cultivate this large field adequately, he should have had the best crop. . . . But in fact he had been too ambitious. So much of the labour of his household was required to pay off the debts (in labour) to those who had helped with the clearing that it proved to be impossible for sufficient credits to be built up for the rest of the season. As a result his field was badly cultivated and the yield per acre poor (p. 72).

Here we find improperly patterned thrusts of energy, with an excess in the clearing and an insufficiency for weeding.

A striking example of deficient input may be seen in Chen Han-seng's *Frontier Land Systems of Southern China* (1949), where he describes an area of Tai-speaking villages in the neighborhood of Cheli, Yunnan Province. Despite favorable climate at relatively low altitudes, ample land, and adequate irrigation, the yields from local rice fields fall well below those of other areas in Yunnan. Table 4.1 brings together data from Lut'sun, a climatically less-favored part of this same province, gathered by Hsiao-tung Fei and Chih-i Chang in *Earthbound China* (1948). To this we have added our data from Bang Chan.
The low yields in the fields of Cheli is evident, less than half of Bang Chan's and a third of Lut'sun's. Some explain low yields on grounds that small areas are inefficient units of production, while others attribute poor yields to diffusing efforts over too large an area as in the just-cited case from Sarawak. As the area of land worked in Cheli is smaller than that in Bang Chan and larger than in Lut'sun, we cannot look to field size for explanation. The important factor is rather insufficient labor, as the column on household size indicates. Cheli has too few workers per house-

Table 4.1. Rice yields, land area under cultivation, and household size in three rice-growing areas

	Rice Yields (short tons/ acre)	Average Area Cultivated per Household (in acres)	Average Number of Persons per Household
Cheli	0.35	4.2	3.9
Lut'sun	1.06	1.1	5.4
Bang Chan	0.86	12.3	5.7

hold. Chen Han-seng amplifies this point, saying that labor is scarce, for extra hired hands are not available, so that only local labor exchange between cultivators supplements the labor needs, redistributing workers but not augmenting them. In addition, many households do not have draft animals and plows, further reducing the energy available per unit of cultivation or at least dislocating it by late plowing with borrowed animals. Considerable areas of land have also been abandoned for lack of population to work it, the result of decades of civil disorders that have scattered people. "The one general, outstanding, obvious and remarkable phenomenon," writes Chen Han-seng, "is its [Cheli's] very low productivity and sparse population (p. 36).

To such cases may be added others that seem to make much the same point—that the energy input per unit of crop must be sustained at a level required for the holders. When the influenza epidemic of 1917 struck India, causing sudden population decline, people did not stretch themselves to plow the old area but reduced the area under cultivation. Certainly the total crop declined, but reducing the size of land that they planted helped prevent further shortages by sustaining the energy expended per land unit. The original ratio of workers to land (0.8 persons to an acre) remained about the same after the epidemic. The Philippine Island of Cebu illustrates excessive energy directed to shifting cultivation; the large number of cultivators, in their search for places to grow a crop, prevents the forests from regenerating in the normal period of years.

With these bits of evidence we find initial support for our statement that an optimal input of energy must be maintained to sustain a given agricultural output. In the following section we shall compare the energy requirements for three modes of cultivation of rice as a step in providing further evidence for this hypothesis.

MODES OF CULTIVATION AND HUMAN ENERGY REQUIREMENTS

In describing the various ecosystems of rice, we have stressed delicacy and precision, as distinct from efficiency or brute power. Though a man can plow a field faster with a tractor than with a buffalo, speed is not necessarily useful. Only where plowing must dovetail nicely with changing weather or where speed is necessary to coordinate one process with another, does the extra energy of a machine become warranted. If plowing can be done at leisure over many days, as in broadcasting, the greater speed of a tractor has no technical justification. Its use resembles flying a jet plane to visit a neighbor who lives only a few miles away. Moreover, the energy expenditure of a tractor must fit the local patterns of labor, lest it become a burden to the owner having to travel miles in search of scarce fuel or a mechanic to make repairs. The more powerful and efficient burst of energy must not interfere with subsequent tasks. In some communities, driving a tractor spoils a worker for the fields, since tractor drivers become too proud to help with transplanting. Like a poor peasant trying to maintain a costly chateau, a simple fisherman operating a battleship, or a crop duster paying the bills for a bomber, the rice grower must avoid being submerged by a machine.

Here we shall compare the quantities of human energy required for each mode of cultivation. Though we shall overlook the niceties of patterned delivery, it should be possible to see indirectly the differing demands made by each ecological holding upon its group of cultivators and to make gross comparisons between them. For this purpose, the natural system of a rice field may be considered as if it were a machine into which man injects a certain quantity of energy and from which he derives a certain output of rice. The input into a rice field comes from the sun, the bacteria of the soil transforming organic material into plant nutrients, and man together with his implements and draft animals. In this composite of forces, we must assume a constancy or only minor variations of the sun's energy and bacteria in order to isolate the human factors of input. Certainly, comparing solely the density of rice plants in a transplanted field with the haphazard stand of a broadcast field, the foregoing assumption is false. The sun's energy contributes more to the thick, transplanted stand than to the broadcast crop, yet it is still possible to isolate partially and compare the human contribution to input. We shall compare the number of days of toil in each field. As for the output of a given field, our measure is the tonnage of rice. The unit for comparison is the growing season for a single crop of rice. The data on yields in Table 4.2 offers our first statement on the outputs of these holdings.

Table 4.2. Modes of cultivation and their yields of rice.

Mode of Cultivation	Yield (short tons/acre)		Number of Cases
	Average	Range	
Shifting	0.645	0.330–1.012	6
Broadcasting	0.594	0.352–2.068	4
Transplanting	0.975	0.220–2.068	7

Sources: See Appendix B.

On the average the output from transplanting exceeds the outputs for shifting cultivation and for broadcasting, which furnishes the lowest yields. The number of cases is small, not because data on yields are scarce but because few yields are clearly associated with any given mode of cultivation.

The range indicates wide variability of yield, considerable overlapping. Here we must bear in mind that everywhere droughts and windstorms may reduce outputs toward zero. The upper limits suggest that broadcasting may exceed the productivity of shifting cultivation, yet we have heard of enormous yields from shifting cultivation in virgin soil that exceed the yields of transplanting. Thus the upper limits of field output by a given mode of cultivation are more variable than the lower limits. We find, moreover, the output of any given rice plot fluctuating broadly, depending not only on the mode of cultivation but also on weather, soil, water, seed, and many other factors.

Some of this variability, however, can be reduced by comparing successive crops on similar or identical fields over three- and four-year periods. Table 4.3 seeks to compare modes of cultivation for this kind of variability. If a given field produces eighty lbs. of paddy one year and one hundred lbs. the next, it would

Table 4.3. Modes of cultivation and variability of yield over successive years in the same field.

Mode of Cultivation	Average Variability (per cent)	Number of Cases
Shifting	54.4	1
Broadcasting	21.0	1
Transplanting	13.5	6

Sources: See Appendix B.

have a variability of 20 per cent, that is, the difference between the maximum and the minimum crop divided by the maximum, then reduced to a percentage of the larger crop. If the crop were identical on two successive years, variability would be zero.

Despite the paucity of cases, our data suggest decreasing variability of output from shifting cultivation to broadcasting to transplanting. As these differing modes of cultivation imply increasing controls over such variables as water level, fertility of seeds, and the spacing of plants, this result is in keeping with our expectations. Of course, conditions are somewhat special in shifting cultivation, for ordinarily these growers change fields each year in order to avoid rapid decline in yield. However, if we may consider that these data and those of Table 4.2 indicate a rough order of productivity, then modes of cultivation can be ranked in output efficacy from shifting cultivation to broadcasting to transplanting. As a system of energy transformation, the output under transplanting is greater than the other two modes of cultivation. We may expect input to follow this order of magnitude also.

Our first indication of energy level on the input side is the population density of regions where these modes of cultivation occur.

Table 4.4 Modes of cultivation and their associated population densities

Mode of Cultivation	Population Density (persons/square mile)		Number of Cases
	Average	Range	
Shifting	31	8–91	6
Broadcasting	255	47–465	2
Transplanting	988	260–1,300	4

Sources: See Appendix B.

More working hands are available as we proceed from shifting cultivation to broadcasting and transplanting. This follows general observation where shifting cultivation occurs chiefly in the lightly populated mountains, and fixed cultivation in the more thickly settled valleys. Yet this does not represent solely a matter of topography, for where population is heavy, as among the

Ifugao and Kalingas of mountainous northern Luzon, the terraced fields of transplanting cover steep mountainsides.

We are accustomed to asking how many people a given acreage supports. Rarely does the obverse thought lead us to ask how many people are required to raise a crop of rice. All modes of cultivation in Southeast Asia share a common feature: that the major portion of the work is carried out by a single household of two, three, or more constant workers supplemented from time to time by the elders, children, and neighbors. Once the land is cleared, each household of shifting cultivators manages its own. A household of broadcasters can do the same until their typically large harvest demands added hands. Transplanting requires workers beyond the household, not only at harvest but also during uprooting and transplanting. The total of these variously patterned inputs in man-days per unit of land per crop season is given in Table 4.5.

Table 4.5. Modes of cultivation and their human labor requirements per unit of land per crop season

Mode of Cultivation	Labor (man-days/acre)		Number of Cases
	Average	Range	
Shifting	67	42–103	2
Broadcasting	17	14–24	2
Transplanting	53	24–133	9

Sources: See Appendix B.

On the average, broadcasters work least and shifting cultivators most. The wide variability, particularly for transplanting, suggests considerable persistent differences from locale to locale, perhaps in such details as the number of plowings and harrowings or the number of times a field is weeded. One upland village may erect fences to keep the cattle from the area under cultivation, but another without cattle need not face this problem. In areas where many predatory birds are living, more time must be taken to tie together a shimmering network of moving objects across the field to frighten the marauders away. Where crabs are likely to snip off a growing rice stalk, more days of watching are required. Thus requirements vary from place to place. In addition we find differences in definition of work. While all observers include the time for plowing, some discount the time for irrigation or for transporting the harvest to the storage bin. None include time for ritual. We expect, nevertheless, that these variations, if corrected, would add or subtract no more than about 10 per cent of the total. If so, the averages may be considered fairly reliable approximations of labor requirements per unit of land for these various modes of cultivation.

The foregoing measures of labor have failed to take into account the area of land utilized in each of these modes of cultivation. The point to note in Table 4.6 on land sizes is that the areas under cultivation by broadcasters are on the average nearly three times as large as those of shifting cultivators. The typical land size of transplanters decreases below that of the broadcasters,

Table 4.6. Modes of cultivation and size of cultivated fields worked by a single household

Mode of Cultivation	Field Size (acres)		Number of Cases
	Average	Range	
Shifting	3.8	1.0– 6.8	6
Broadcasting	10.5	8.0–12.3	2
Transplanting	5.5	2.0–12.3	6

Sources: See Appendix B.

but the smallest plots are those of shifting cultivators. The narrow range for shifting cultivation reflects the limitations of man power and tools that make preparation of large plots difficult. The broadcaster with his buffalo and plow can manage larger areas and larger harvests, while population pressures may limit the availability of land for transplanters.

We are now in a position to approximate the direct labor input for a given crop raised by a particular mode of cultivation. The *direct* labor input refers to labor on the immediate crop (plowing, seeding, weeding, and so forth) as distinct from indirect work such as preparation of tools, caring for animals, and maintaining the irrigation system. By multiplying the rate of labor input per unit of land by the land area used, we have a rough measure of the energy needed to utilize a given mode of cultivation. In Table 4.7, we have multiplied the average man-days necessary to cultivate a single acre (from Table 4.5) by the average number of acres used for a particular mode of cultivation (from Table 4.6). So viewed, the least investment of energy occurs among the broadcasters, the greatest among the transplanters.

Table 4.7. Modes of cultivation and their direct man power requirements per crop season

Mode of Cultivation	Man Power Rate (man-days/acre)	Field Size (acres)	Man Power/ Crop Season
Shifting	67	3.8	255
Broadcasting	17	10.5	179
Transplanting	53	5.5	292

Tallying the indirect man power requirements is somewhat more difficult, for one must answer such questions as the following: Is the making of agricultural tools a part of production? Must the cost of land and draft animals also be included? If so, must one tally the hours of feeding and tending the animal throughout the year? Should the cost of the shelters that cover both man and draft animal become a part of the indirect man power requirements? Must we also count the energy expended in raising children to be growers? Ideally, one might amortize each year all the past inputs in the history of a social group, insofar as a tangible residual remains, adding some fraction of the energies expended to repair a shed, recover a roof, dig out an irrigation canal, raise a child to the age of being a productive worker, maintain him through each productive year, and bury or cremate him with appropriate ceremony. Such inputs on which other inputs depend might be fractioned according to time devoted to raising a crop of rice, but the parameters of definitions have extended too far for easy analysis, barely to mention devising convincing criteria for classification and quantification. Hence we have chosen to draw the curtain once the expenditures for tools, land, draft animals, and irrigation systems have been entered.

We have Janlekha's data from Bang Chan showing the minimum equipment needed for transplanting (see Appendix B). It includes shovels, pitchforks, boats, and such devices as a windmill and dragon-bone pump for irrigation. Purchased in the market at 1948 prices and translated into U.S. dollars at the rate of $1 to 20 *baht*, a transplanter's total equipment costs $200, which may be amortized according to the life of the particular item, between 1 and 20 years, at $24.65 per year. Similarly the two buffaloes with a life expectancy of 10 years and an initial cost of $50 each may be amortized at $5 each per year. Finally we enter land values as if they were rented at the going rate for transplanters, $6.25 per acre. For broadcasting, the equipment has been reduced by omitting items referring to irrigation and lowering rents (because of diminished unit production) to $3.20 per acre. Since shifting cultivators pay no rents, utilize no draft animals, and work with a limited range of hand tools, their amortization requirements are proportionately reduced. These materials follow in Table 4.8, with rents based on land size from Table 4.6. Here we have both calculated the "capital" cost of each mode of cultivation and then converted these figures into units of man power. For this conversion our rate of $0.50 per day is near the average wage for agricultural workers, perhaps a little high for the year 1948 and a little low for 1970. The indirect man power requirements for shifting cultivation stand in sharp contrast to the heavy ones for broadcasting and transplanting. While the tools and equipment of transplanting are more demanding, the costs are nearly the same because of renting larger areas for cultivation by the broadcasters.

Table 4.8. Modes of cultivation and their indirect man power requirements per crop season (represented as $U.S.)

	Mode of Cultivation		
Item of Cost	Shifting	Broadcasting	Transplanting
Tools and Equipment	$2.00	$17.50	$24.50
Land Rental	0	$33.60 (10.5 acres)	$34.38 (5.5 acres)
Draft Animals	0	$10.00	$10.00
Total	$2.00	$61.10	$68.88
Man-day equivalent at $0.50/day	4	122	138

One significant factor has, nevertheless, been omitted, and this forms one of the most critical differences between broadcasting and transplanting. Here we refer to water control. Though our estimates of man power input include the labor for running water into fields from existing channels, they fail to hint at the underlying inputs. Just to level and dyke a field for transplanting, such that the water stands evenly on the surface of the ground, may well require more than 200 man-days. And how long will it take to amortize it? Twenty years? A century? Ten centuries? More important, the digging and maintenance of an irrigation system may be exceedingly costly. A simple ditch provides water for most villages in the Mae Ping valley of northern Thailand, the maintenance and amortization of which may require more than one hundred thirty man-days per mile per year. Hydraulic systems with reservoirs and dams vary even more, from an estimated $2,000 for a small concrete dam to some $12 billion estimated for the projected Pa Mong Dam on the Mekong River. Such variation plus the uncertainties concerning the length of time needed to amortize the costs leaves this investment in irrigation too varied to count as more than an uncertain factor of X. Bang Chan, for example, has easy access to water, thanks to the water gates at Rangsit and the Pratunam, erected by the Thai government about 1907 to stabilize the water level. The canals that lead the water to the various fields must be dredged anew at intervals of about 50 years. In view of these variations, to assign a cost of amortization would at best be arbitrary. Whatever its size may be, it must be added to the input requirements for transplanters.

It is now possible to compare inputs required by various modes

Table 4.9. Total human labor input for growing a single rice crop under three modes of cultivation

(represented in man-days)

Type of Man Power Requirement	Mode of Cultivation		
	Shifting	Broadcasting	Transplanting
Direct	241	179	292
Indirect	4	122	$138 + X$
Total	245	301	$430 + X$

of cultivation. In Table 4.9, the figures on direct man power requirement come from Table 4.7, the indirect ones from Table 4.8. The *X* is the unknown long-term investment needed for irrigation systems. Here are approximations of the human input requirements for raising an average crop of rice by each of these three modes of cultivation. On this basis transplanting requires a relatively high input, more than a third above that needed for broadcasting. The input for broadcasting, in turn, stands about one-fifth higher than the input needed for shifting cultivation.

Our thesis showing that increased energy is required for growing a single crop of rice by these increasingly complex modes of cultivation reflects the findings of Boserup's *The Conditions of Agricultural Growth* (1965). The author observes that "a certain density of population is a precondition for the introduction of given techniques" (p. 65). Her view as well as ours reverses the thesis that underemployment is rife in Asia and that these countries suffer from the Malthusian blight, too low an agricultural productivity, and too heavy a population. Along with Myrdal she indicates that, even in densely populated India, people lie no more heavily on the land than in countries such as England or the Netherlands. Poverty results from malutilization of labor, rather than overpopulation: "[In India] the labor force has been utilized, but not as intensively or effectively as it might have been. Consequently, average indexes of yield, whether per unit of cultivated land or per worker in agriculture, have remained low. That considerably higher yields could be obtained through increased intensity and efficiency of work is proved by the example of Japan and many areas in South Asia itself" [Gunnar Myrdal, *Asian Drama* (New York, 1968), p. 436]. Here we have given ecological and anthropological phrasing to this view in saying that the labor input of a given social organization must match within tolerable limits the input requirements of its holding. In Japan, but not in India, the complexity and hence labor requirements of holdings have increased by using more demanding modes of cultivation.

To observe the relation of input to output in these modes of cultivation, we have constructed Table 4.10, showing input in man-days (from Table 4.9) and output as short tons of rice. Our figures on output are derived from the yield for each mode of cultivation (from Table 4.2) multiplied by the average size of field for each mode of cultivation (from Table 4.6). The ratio of man-days to crops appears in the final column.

The ratios of output to input show that per unit of work more rice is raised by broadcasting than by the other modes of cultivation. A day's work at broadcasting nets almost twice as much grain as a day's work at transplanting and nearly three times as much as by shifting cultivation. Though other samplings of outputs and inputs may alter these ratios somewhat, broadcasting probably will remain the most effective producer per unit of human labor.

Table 4.10. Relationship of input to output in three modes of cultivation during a single crop season

Mode of Cultivation	Input (man days/ crop season)	Output (short tons of rice)	Output/ Input × 1000
Shifting	245	2.451	9.08
Broadcasting	301	6.237	20.39
Transplanting	430	5.363	12.47

What do these ratios tell us about the varying modes of cultivation? The broadcasters get more returns for their labor, but we may not make direct comparisons of efficiency on this basis. To do so requires converting output and input into common units. If we may say that for each day of work 1.1 lbs. of rice are expended, this average adult dietary requirement in rice may then be compared with the size of the crop. Thus in shifting cultivation, the worker puts in 1.1 lbs. of rice per day and receives 17.6 lbs. Using the standard formula for efficiency (output/input × 100), shifting cultivation has an efficiency of 1700 per cent, broadcasting 3800 per cent, transplanting 2300 per cent. Few machines work as well, but few receive as much support from sun and bacteria.

We must, however, resist using this basis of comparison as the sole criterion of value. Even comparing efficiency of automobile engines quickly passes beyond efficiency as soon as conversation turns from ordinary highways to facility in climbing steep grades or fording streams. Then power and traction become important. Similarly, modes of cultivation, in addition to feeding people, must fit other specifications, among them the topography, the water resources, and the population density of a given area. The value of a mode of cultivation depends on many factors, and the civilization resting its decisions on any single criterion, whether efficiency, monetary profit, or the programmed formulation of a computer, is heading for trouble.

Instead of regarding these modes of cultivation as crude or refined, developed or underdeveloped, let us see them simply as the means that help provide man with his supply of rice. Shifting cultivation, producing adequately but not excessively, fits well a scene with multiple resources to sustain a light population. In the forests are alternate foods to gather and animals to hunt. Labor requirements for cultivation rest lightly on the workers, so that they may turn toward other pursuits to sustain them when the crop fails. The system is profligate of land, but in areas of limited population density, this offers no limitation. Requiring few capital tools, cultivation is available to many, and the supporting society may remain relatively undifferentiated. Shifting cultivation offers an accessible tool that produces modest returns.

Broadcasting demands more labor but over the years less land, returns a more abundant crop, and can offer it to a larger population. Man must in return spend more days of his year at cultivation and depend more on his field for sustenance. His surer crop draws him away from hunting to domestic pursuits, to making plows, caring for livestock, and weaving baskets. He is limited by the availability of suitable lands and the necessity for considerable accumulation of capital. This extensive mode of cultivation is like that of the Kansas wheat fields, plowed, seeded, and cultivated by mechanical means. Given a comparable shortage of man power plus an insatiable demand for food, many areas of the Orient might have become mechanized like the Great Plains of North America.

Instead, populations increased and land became scarcer, so that transplanting became a more suitable mode of cultivation. The heavy yield and labor requirements make the fit admirable. Becoming a cultivator of rice fields narrows activities to skillful refinements in managing plants and water. The diet of rice flavored with vegetables and a little fish contrasts with the varied meats, roots, and nuts of the shifting cultivator. Even costumes have changed; austere black pajamas have replaced garments decorated with seed necklaces and fur. The households of workers have become specialized parts of society which cannot sustain themselves and must be protected by a larger society.

Rather than stages in development or evolution, these modes of cultivation are manners of adapting to a changing environment. The increasing populations offer a greater work force to sustain a more demanding ecological holding, yet this is not a one-way road. A lone widow in Bang Chan raised a crop of rice with a digging stick, while her neighbors with many children continued to transplant. Indeed, the ones who stick to their techniques like the underpopulated village in Yunnan mentioned earlier in this chapter may in some way have failed—or had we better regard them as the stout-hearted ones who would not give up? The direction of evolution is more easily read long after the event. Now we can better marvel that the sequence of gatherers did not extinguish all the world's rice but somehow found ways through planting to increase and stabilize a food resource. When land became scarce, by turning to plows and buffaloes in the soggy wetlands, they grew larger crops for more people since rice, unlike wheat, can tolerate water. Then by controlling the water, others reached a new accord with their scene on the higher lands. At each moment of distress, someone managed to comprehend nature's response to their insistence and reply suitably.

Finally let us return to our generalization concerning optimal labor input for any given holding. Certain direct evidence for the validity of this proposition rests on "clinical" observations of failing communities of rice cultivators. However, the unique labor input requirements for each mode of cultivation and the varying but associated maximum crops, each rather distinct, are consistent with its implications.

RECAPITULATION

Before proceeding to the societal changes in Bang Chan that accompany changes in the mode of cultivation, let us restate briefly the formulations of the last two chapters. They may be written in prose much as one summarizes the tale of an opera, but let us try for greater lucidity and succinctness than are found in most résumés. We shall use both a mathematical statement and the most abridged form of poetry that we know, namely a haiku-like statement. Both convey complex ideas simply. The two kinds of summaries differ in their mode of forming a symbol. A mathematical symbol raises the level of generalization through greater abstraction, as in moving from *hawk, ostrich,* and *wren* to *bird,* thereby eliminating all but the essentials. A poetic symbol often reduces the level of abstraction toward greater concreteness. The resulting image also stands for the whole, as in reducing the foregoing collection of birds to a "muttering of horny beaks." A further contrast arises: in mathematics the forming of an equation consists in balancing symbols about the fixed "equals" sign; in poetry the making of a haiku-like statement, after the relevant symbols have been found, calls for selecting appropriate middle terms to join the symbols. Thus both mathematics and poetry should enable us to recapitulate briefly and lucidly.

Having surveyed the various sources of energy needed to raise a crop of rice, we have sought to isolate and define the societal contribution. This may first be approximated with three variables: (1) the number of people in the work force, which may be represented by p; (2) the duration of the work through the crop season, represented by t; (3) the area of land to be worked, represented by L. Using these generalized symbols, we reach the following mathematical statement, where f is the symbol for *function.* We may translate it to read "is a function of," a kind of shorthand indicating that further definition is needed to make the equation precise. Our statement then reads:

$$\text{Societal input of energy} = f\,\frac{pt}{L}$$

In addition, we have observed that the input of energy also depends on two other factors: (4) the cultural definition of workers and work days together with its prescriptions for industry, work quality, and perseverance; this we represent by C. (5) The particular mode of cultivation that is employed, represented by M. Joining these two additional terms to the mathematical statement, we write:

$$\text{Societal input of energy} = f\,\frac{CMpt}{L}$$

Thereby it may be said: The societal input of energy into a given ecological holding is a function of the social organization and the mode of cultivation by a particular working unit of population tilling throughout a given crop season on fields of a certain size. Yet such a generalization obscures our major point: Each mode of cultivation is appropriate to a scene of varying population with differing economic command and varying availability of land. While shifting cultivation best fits among scattered people, transplanting presumes many people working small holdings, while broadcasting forms an intermediate step between the two.

Somewhat the same statement can also be made in far fewer syllables:

SHIFTING CULTIVATION

Sun again; the rain
Erased my foot print
In the powdered ash.

BROADCASTING

The gray rising flood
Has smoothed
The furrows of my plow.

TRANSPLANTING

From dyked fields
Empty straws have sent
Their tribute to our village.

THEIR SEPARATE WAYS

Eyes of plowmen
Cannot see clouds
Wedged between the hills.

Part II

ADAPTATIONS TO A CHANGING ENVIRONMENT

In Part II we return to Bang Chan. Here through a dozen decades people have been transforming a savannahlike wilderness to a land of rice fields, so altering untold ecosystems of the region. In the course of these changes the inhabitants grew rice first as shifting cultivators, then as broadcasters, more recently as transplanters. The task may be stated simply as adding man power and tools to productive units, but like grafting bones or replacing organs, the adding of living members to functioning organized groups may result in either debilitating rejection or increased vitality for the group. Studying these people and their social arrangements, we shall seek to determine how their total living changed as they adopted each new mode of cultivation. By pulling together portions of what contemporary residents have said about themselves and their grandparents, by combining these fragments with observations on still existing pioneer settlements in other parts of Thailand, we shall fashion a sequence of portraits of Bang Chan at four periods of its brief history. Within the limits of our understanding these changes will be treated as adaptations to a changing environment.

Let us make clear that in speaking of adaptations we distinguish the concept from development and social evolution. Adaptation is humbler, dealing with such miniscule events as the prehensile shoots from vines, the molar teeth of vertebrates, and the increase of polygyny in a social group. Whether such adaptations can be called a contribution to some all-pervasive trend poses questions concerning development and social evolution. They stand beyond the scope of our present work. Of course, evolution may be found in the succession of modes of cultivating rice, for each change depends upon its predecessor. Before broadcasting could occur, agriculture had to be discovered; similarly, transplanting depended on finding that rice could survive in water. These technical sequences are of less interest to our present concerns than the way societies utilize these inventions. After all, transplanting, the last of the three modes of cultivation to be invented, was probably used in China about the time of the birth of Christ. Archeologists have found in Southeast Asia traces of dyked rice fields in the neighborhood of the ancient capital of Burma in Pagan; the city dates back to A.D. 1000. In the Red River delta of Vietnam transplanting seems to date back that far also. These techniques were known in Southeast Asia, yet the dyked fields in Pagan were abandoned for broadcasting, while those of Vietnam continue in use to the present day. By regarding each case in relation to its environment, we may understand certain kinds of societal changes without raising the additional question whether Burmese adaptations were retrogressive and Vietnamese, progressive.

What may be called societal adaptations are exemplified in the following quotation, the observations of Nai Sin Saensug, a Bang Chan man who wrote a brief history of his locality:

> The way they grew rice [formerly] was very primitive. They had no buffaloes and no tools such as sickles and windmills. They used a mattock to loosen the soil. They used a stick to make holes in which they put five or six grains of rice and covered the holes with soil. They did this in the sixth month [probably June] until their rice fields were fully sowed. When it rained, the rice sprouted. During this time they had to kill weeds very often until the sprouts grew into clumps which the weeds could not overcome. . . . If they had buffaloes, they plowed their dry land and sowed seeds and waited for rain. When it rained, rice and weeds both grew. So farmers had to kill weeds nearly every day.
>
> The broadcasting method of cultivation yielded less than the primitive digging-stick method for equal amounts of land. But they could grow more rice by broadcasting more land. So in ancient times they grew 2 or 3 *rai* [of rice] for a small family and no more than 10

for a big family. But when they had buffalo and changed to broadcasting, they could cultivate 100 *rai* or more depending on the size of the family and the number of buffalo.

The development of the irrigation equipment during the past 50 years began by using buckets and human energy. The water shovel came after this and then the splash wheel [instruments for flooding a paddy field]. These two devices still require human energy. When windmills were invented, human energy was less used. Nowadays the windmills are used everywhere, but farmers prefer engines if they have enough money to buy them.

At first glance this statement appears as utopian as those of the most Darwinistic historian of the nineteenth century. "Primitive" is clearly associated with historically earlier features, and the more recent ones, though not pronounced "modern" or "advanced," often seem to assume progress. Yet the familiar occidental phrasing of these changes does not use such conditional sentences as "If they had buffaloes, they plowed," or "Farmers prefer engines, if they have enough money to buy them." Even allowing for slips and falterings, progress and evolution rarely receive conditional phrasing. Instead, the author of the passage seems to be dealing with a series of adaptations. Having mentioned the technical advantage of plow and buffalo, he stresses the importance of a small or large group of workers to produce a larger crop. The mechanical advantages of the various techniques of irrigation are less important than whether they depend for their operation on the strength of human beings. Between windmill and engine for powering irrigation, he raises no question of mechanical advantage, merely the grower's preference. So we seem to be dealing with the life-supporting adjustments between a social group and its environment. They may cumulate to form a trend, or what is done this year may be abandoned the next. For a moment, nevertheless, they change societal being.

Chapter 5

YEARS OF SHIFTING CULTIVATION: 1850–1890

At best we can date Bang Chan's beginnings within the decade 1845 to 1855. There are no records, thanks to mildew, centipedes, and termites with their appetite for paper. Even if these documents could have been preserved, no local records go back that far, for in 1850, government did not exist in the area. No living person has resided in Bang Chan for more than 90 years, and if fathers or grandfathers told tales of pioneer days, they had been forgotten when we enquired during the 1950s. Our next best reference is the date of finishing the Saen Saeb Canal because it lies in the vicinity of Bang Chan, yet the references also present problems, despite clear mention in the dated chronicles of the time.

Saen Saeb Canal was one of King Rama III's (1824–1851) responses to a revived dynastic vigor in Vietnam. This new force threatened the Thai kingdom's eastern network of alliances with Cambodia, Savannaket in Laos, and other Mekong River city states. Through a canal cut in a northeasterly direction from Bangkok to the Bang Pakhong River, an army might reach Battembang at least two or three days sooner, possibly a week in case of storms, than by paddling through the Gulf of Thailand and entering the Bang Pakhong River at its mouth. So Chinese were hired to dig with mattocks and to move the earth by baskets from a proposed ditch with a depth of 10 feet, a width of 40 feet, and length of 33.2 miles. Chronicles of the reign tell of the king giving this order to a minister in either 1832 or 1837, depending on the edition that has been consulted, but both editions agree that the work required 13 years to complete. Thus with Saen Saeb Canal completed in either 1845 or 1850, people could not have settled in Bang Chan much before those dates. For purposes of settlement the area was inaccessible and uninhabitable by Thai until the canal had been filled with water.

The Thai refer to a savannahlike area by the word *tung*, ordinarily a vast flat space covered with tall grasses, sedges, and in this region, on the higher points, with thick bushes. Accustomed to living on river banks, where water for the household, fishing, and transport was readily available, Thai avoided these areas. Through the rains the grasses blocked travel by boat, and on foot during the dry season one easily lost the way. Even as late as the 1890s, Bang Chan migrants reported that when they were mounted on the backs of water buffaloes, only grass was visible to right and left. Besides, snakes "as big around as an arm" slithered in the thickets; occasional crocodiles lurked near permanent ponds. However, these creatures were not as fearsome as the elephant herds that might charge and kill the man who surprised them. In the proper season, deer ranged searching for tender shoots near water holes, and following them the tiger. A few Khmer-speaking hunters, who settled long enough to grow patches of rice, moved about in the great expanse, but their groups probably rarely exceeded 10 to 15 households as compared with the hundreds in Thai villages that dotted the shores of the Chao Phraya River.

When some Chinese coolie had carried the final basket of alluvial soil from the oozing bottom of the great ditch, and the barriers holding back the water had been pierced to fill the Saen Saeb Canal, a thin new strip of land had been made habitable. About this time ministers of state and victorious generals were rewarded; an appreciative monarch granted them lands along the canal banks. Though little interested in living there, these men settled their newly acquired lands with the less highly skilled prisoners of war, who were then expected to support themselves on garden patches with seed and a few chickens provided by their masters. These prisoners of war, captured on past campaigns against Malay sultans and Laotian princes, were additional tokens of appreciation for service to the crown. Assigned to the households of worthy notables, over the years their numbers had increased. The new land on the Saen Saeb Canal provided a welcome outlet for overmanned houses, gardens, and elephant stables near Bangkok.

In 1860, a certain D. O. King, Esq., of Newport, R.I., reported to the Royal Geographic Society of London his travels by boat on Saen Saeb Canal from Bangkok to Cambodia:

> This canal is 55 miles long and connects the city of Bangkok with the Pang Pa Kong River and is made through a flat alluvial country devoted entirely to the culture of rice. The natives, like the rest of the Siamese, appear to be a branch of the Malay family. The floors of their bamboo-thatched houses are raised some 4 feet from the ground; their clothing is simply a cloth around the waist; and, whatever they may be engaged in, one hand is generally actively employed warring against the swarms of mosquitoes (pp. 177–178).

This traveler certainly miscalculated the length of the canal and in all probability the depth of bordering rice cultivation, yet his report of the monotonous clusters of bamboo huts on both sides of the sluggish length, all whining with mosquitoes, agrees with the independent description of our informants. Beyond the sight of this traveler, who sat under a rattan awning at the bottom of a sampan, lay the broad shadeless savannah. About this we were told by a woman of 90, nearly senile and bent by age to a right angle when she walked. Her father, newly wed and childless, came to Bang Chan in the early 1850s.

THE FIRST MIGRANTS

Chaem, father of our 90-year-old informant, came from a village on the Chao Phraya River upstream from the spires of the king's palace in Bangkok. This village, during the mid-nineteenth century, was a port for boat traffic up the river, and was settled mainly by Chinese from Canton and Hainan who married Thai women. The old woman in Bang Chan never mentioned her father's occupation in this small port, though she did say he was one of three brothers and one sister in her grandfather's household. Her father may have been a stevedore carrying heavy bundles from the boats into some company shed, a sailor on one of the boats, or perhaps just a gardener who supplied green vegetables to the wealthy owners of mercantile establishments. At his death he left a wife, a daughter married to a prospering merchant, and three married sons. Without the particular liaisons of their father in a community tightly controlled by the local guild of merchants, and seemingly at odds with their sister's husband, the three brothers found themselves alienated and unsheltered. Their sins we shall never know, though a community of this sort could often tolerate a thief, sometimes an assassin, and could always bind a man to work off his debts. In any case, Chaem and his two older brothers with their widowed mother and their three wives, one of whom brought her mother too—in all a party of eight plus perhaps two or three children—set out in three boats for the newly accessible lands along the Saen Saeb Canal.

Squatters could scarcely expect to settle the already claimed lands on the canal banks, and so the three brothers headed their boats up one of the natural creeks feeding water into the new canal. They were well advised, since permanent water is indispensable for a household. For a time the channel continued between great clumps of grass where straying into a blind alley was easy. When the winding channel ended, a person wading in the black muck might blindly push his boat some hundred yards or more along the track of greatest moisture, until the moisture itself disappeared. Then a person determined to proceed still farther had to dig his own channel, for over the centuries the stream, choked with rotting foliage, had forced a subterranean channel. Black, humus-rich soil furnished the spoor. A single penetration by boat in a narrowly dug channel frequently sufficed to draw the keystone, for then the rains washed the loose vegetable debris down to the adherent gray-brown clay. Thus nature dug many of the present canals across the savannah, among them Bang Chan Canal, which then extended a little farther as each new migrant mattocked his channel a little deeper into the thicket. On the map (see Map 3), meanders easily distinguish the natural water courses from the geometric canals dug by man.

At about the point where Bang Chan Canal turns westward and Khib Mu Branch runs northward (see Map 4), the three brothers chose to build their house. As Bangkok means "village of the clump of grass," Bang Chan means "village of an elevation," and those newcomers may have selected some long-since-disappeared but then perceptible elevation to raise them a little higher above the inevitable floods of the rainy season. Whatever may be said of this lonely spot immured in tall grasses, the land was virgin and uncoveted by any other person. Water was accessible in all seasons. About them stood enough bamboo for house posts; other pieces could be split for mats, walls, and floors. At their elbows grew grass enough for a hundred roofs, and a house could be completed in a day.

The land promised nothing but struggle. Acccording to Buddhist doctrine only people who have seriously sinned are punished, and these new migrants were sentenced by the moral law of natural justice, the Buddhist law of Karma, to life terms at hard labor for some known and many forgotten crimes from past existences. Of course, some say today that laboring for a living in the raw savannahs is no more severe than a coolie's life in a city, but all agree that the uniform dreariness of the wilderness is far more depressing. The variations are of season only, not of person, scene, or event—no stopping for a nip of liquor with a friend, no boat races on the river, no friendly gambling at New Year. Instead, there is isolation, with few occasions for gaiety, and the foreboding blackness of night outside the bamboo shack.

Just to feed a household requires many hours of toil. If studies made in the 1950s apply, the average meal required at least two hours of preparation, meaning time to clean the fish, peel the vegetables, pound the condiments, toss the rice free of dust, and kindle the fire. Like contemporary cultivators, the pioneers cooked but twice a day, though they doubtless snacked on leftovers at midday. To these four hours of preparation must be added at least two more hours to gather the herbs, seeds, greens, and roots, all locally available to those who knew where to look. In a few years the newcomers learned where to find them.

Each hour of food preparation competed with time needed for housekeeping tasks. Fortunately, the newcomers came with fire-making equipment and pots for cooking, but every other life-sustaining occupation required hours of work. Capital equipment for fishing, hunting, or cultivating a garden had to be fashioned before the fruits of labor could even be sought. They soon needed a rat-resisting storage space for grain, a mortar and pestle for milling, trays for winnowing, bamboo tubes for storing water, mats for sleeping, blocks for head rests, hats to shade the midday sun. Yet all the while they had to repair the leaking roof, maintain the aging boat, sharpen the dulling tools, mend the ripped loin cloth. These chronic demands were merely augmented when a woman was pregnant and could do but half her customary chores, or when a man was flat on his back on a mat, his teeth chattering with fever, or when a white-haired elder could no longer regather strength to fetch one more bamboo tube of water or one more bundle of fuel for the fire.

Within this scene of expiating sins, fishing furnished a clear exception to the general rule of arduousness. In these channels newly opened by human migrants to light and air, fish suddenly swarmed. The typical native species in this setting is plachon (*ophicephalus sp.*), which burrows deep into the moist clays to sustain itself through the period of dessication and then pops up, like a comic character, when good wet days are back again. The major gain in fish, however, comes from species requiring permanent water, so neatly provided by Saen Saeb Canal and its tributaries. So important did fishing become in the region that the major settlement a few miles east of Bang Chan was called "Fishtown," just as Leadville and Silver City were named after their chief products. We shall know Fishtown better by its Thai name of Minburi, where *min* is the elegant Sanskrit word for *fish* and *buri* the Sanskrit postfix for *town*, more easily recognized in its English form as *burg* or *burgh* (Pittsburgh) or its Hindu form *pur* or *pore* (Singapore). Minburi became a trading center for fish, then headquarters for district government, and subsequently a rice-milling center.

In those days fish abounded, but to catch them still required an investment in the proper equipment for the season. In deep water a man paddles at night. On the bow is a friend who stands poised with a flaming torch in one hand, a spear in the other. At midlevels one may use nets fixed to a boom that can be raised or lowered even by a child. With just a few inches of water over the mud, the fisherman walks into the pools and jams one end of a cylindrically shaped basket around a fish; then he reaches in and pulls the fish out through the open end.

Chaem and his brothers found it most convenient to sell dried fish in Bangkok for the indispensable staples: their tools, pots, salt, and cloth. In the cool season of diminishing waters, two weeks of fishing produced more fish than could comfortably be sent to market in small boats. The newly-caught fish were cleaned; then the meat spread on trays in the sun, turned from time to time for proper curing, and finally packed into baskets or mats for transport. Though hundreds of others must have been carrying out these same motions all along the banks and tributaries of Saen Saeb Canal, the market for preserved fish seems to have been both stable and unsatiated.

Limited to hand tools, the pioneers grew their rice by shifting cultivation or at least applied these techniques to a single patch of ground. Brush and grass could be quickly cut with a stout knife and burned in less than a month of dessication. Chaem with three adults in his household needed about 2 *rai* (0.8 acre) of cleared land to feed his household. If the virgin soil actually yielded 40 *tang* per *rai,* as contemporary informants maintain, 2 *rai* would have grown 1,936 lbs. of unhusked rice, an abundant supply for three people for one year. A third *rai* might have provided a surplus, but the added work would overtax the group.

While fishing is limited mainly by the endurance and perspicacity of a fisherman, rice cultivation concerns many jurisdictions. Before cutting the soil with a mattock or a plow, one should first have offered food and flowers plus a burning incense stick (if available) to the Lord of the Place. He insures the growth of the plants. Yet this simple ceremony without the preceding royal plowing rites at the capital may well be ineffectual. The king in his capacity as Lord of the Flatness of the Earth addresses higher beings in the hierarchy of gods and angels: Mae Thorani, goddess of vegetation, Phra Phum, Lord of the Earth as a whole, and perhaps also Nang Megtala who brings the rain. With word passed down from on high, the many local guardians are prepared to assist in every valley and backwater.

As we have seen, two household members alone can carry on shifting cultivation for most of a season. A man drills the holes for planting; a woman follows, dropping seeds into the holes. The rice emerges as clumps, which must be freed of competing weeds until the rice plants have established themselves. Then is time to clear an extra patch for sugar cane, eggplant, squash, and beans.

Critical junctures appear in the normal growing season. Since rice (like human beings, elephants, and water buffaloes) has a soul, it requires special treatment during its period of gestation. The rice plant has sprouted from the womb made in the earth by the digging stick, has sprung from that womb to become a plant, and at the moment of budding must receive the treatment accorded a pregnant woman. She craves bitter tastes—limes or lemons; she stands at the summit of her beauty and so enjoys seeing herself in a mirror, as well as smelling perfume and scented powder. Cultivators are happy to offer such comforts to please her, but as she is shy as well as ravishingly beautiful, only women may carry these offerings to her in the fields. A man, unable to withstand the attraction of her beauty, would frighten her away, and hence ruin the crop, as tales of legendary times have long demonstrated.

With the grains fully ripe at harvest time, one should next ask pardon and give thanks to Mae Phosop, the Rice Mother, but the approaching reapers have already frightened her away. The rice thus cut and carried to the threshing floor has lost its soul, and only a woman can recover it. Amidst the stubble left by the reaper she gathers some of the rice grains left in the field, placing them in a little basket. Then from the rice straw she makes a doll, a span of the hand or so in length, and having nestled it among the rice grains in the basket, she softly calls the soul of the rice to come and enter the doll. Covering the whole with a cloth, she carefully carries it to the household, as if here were a royal princess journeying in a palanquin, spared from the prying eyes of passersby. A few hours later the soul of the rice is ceremonially installed in the granary, thus insuring fertile seed for the coming year.

Were these rites all smoothly accomplished, the season was good, yet further dangers to cultivators in Bang Chan might appear in another quarter. While ripening grain bowed the stalks more sharply each day, a passing herd of elephants could in less than five minutes mash the field flat and the houses of the three brothers too. Even distant trumpeting alarmed the little settlement. At that moment he who happened to know the efficaciously polite manner of addressing the leader of the elephants called out toward him in tones gauged to awaken pity and turn the herd in another direction. Of course, Mae Phosop had been asked, when given the bitter fruit and sweet-smelling powder, to guard her children against destruction, but her best efforts needed reinforcement. A wild pig might intrude at night; swarms of rats sometimes descended like a flood on the ripening grain; flocks of ravenous birds were waiting in the surrounding bush for the quiet moment to swoop in. So the last weeks of the crop required continuous vigilance, someone to guard the field day and night, others to suspend twisting and darting objects over the grain to hold the birds at bay, children with slingshots and mud pellets to chase away the boldest ones. Despite the best efforts of people to abet Mae Phosop, many times the granaries were little more than half-filled when all the crop was in.

Each of the three households performed most of these tasks alone, yet they sometimes joined together to make living easier and a little less lonely. Jointly they built the three bamboo huts and in those first years might well have helped each other in cutting and burning the separate fields. Occasionally one spared another's efforts by selling the other's fish in the Bangkok market. The job of turning over a ponderous dug-out boat for repairs was expedited when three people worked together. In proper season, two men hunted, and the sounds of night were less frightening when those left behind slept together in a single dwelling. Then a feast of wild pork or venison marked the hunters' return, and oft-told tales of shooting with the crossbow kept the group gaily together into the night. Doubtless the women helped each other as midwives during birth; children were left with an old woman while mothers weeded the fields. Together they buried the dead, led by the one who could remember some phrases of the formula that prevents the dead from mingling with the living.

Here was less a collectivity than three adjacent households. They worked no joint patches of rice, each household dried and ordinarily sold its own fish, and each cooked its meals under its own roof for its own members. One may have loaned or given food to another in time of emergency, but a return of some kind was expected. Had a need for common action arisen, the eldest might have helped implement a joint effort with some little margin of the authority accorded to age, yet in their isolation no such event seems to have arisen that drew the three brothers together. What lies behind this loose federation? Why was there no more organization than this?

THE ORDERING OF A KINSHIP COMMUNITY

As the Thai build their simple bamboo houses so that they can be readily modified and quickly dismantled, so too they build their groups. Because of the continual changing of the membership living in a single dwelling, we shall translate the Thai word *khraub khrua* (literally, "cover of a hearth") as "household" rather than "family," to avoid suggesting a limited and fairly stable group of people. Each generation calls out the particular personnel for its own household and expects the group to be so rickety after a generation of joint living that a successor must rebuild entirely. This rebuilding in each generation characterizes the kinship systems that Murdock (1960), among others, calls "cognatic" (or bilateral), in contrast to "lineal" systems where social architects have organized groups that survive several generations of buffeting. In Thailand, a family line is missing, though Rama VI (1910–1925), after years of schooling in England, proclaimed that each family in his kingdom should take a patrilineal surname. People sought out names by one or another means, yet the plan succeeded only in part because children often changed the family name, obliterating the connection between generations. Sons, moreover, may move to the bride's household when they marry, so that continuity of generations can run more consistently between mother and daughter than between father and son. At death a man tends to dissipate the body of his estate by dividing it among his daughters as well as sons. In the household of a cultivator there is precedent for giving the house to the youngest daughter along with a portion of the cultivated fields, for she with her husband may have cared for the old man or woman through the final years. Nevertheless, he is quite free to bypass all of his children. Even in the case of the Thai throne itself, the successor is not necessarily the eldest son of the eldest queen but is chosen in council as the ablest among all sons of the king, so even newly-crowned kings must build anew.

Certainly the easiest way to gain a helper in a household, and often the cheapest, is through marriage. A poor man simply persuades some girl to elope with him, and these little dramas are often performed. At the cry of an owl outside in the dark, the prospective bride declares to her parents that she is going to relieve herself in the garden before retiring. In the darkness she meets her lover and his confederate who has made the final arrangements, and down the path they run to a waiting boat, thence along the canal to shelter in the household of some friend of the groom. Some parents are said to rush out at the hint of disturbance, but ordinarily they arrive too late to intervene. Their daughter is gone. Commentators have sometimes found parallels between this practice and marriage by capture, yet local jokesters like to cite the cases where eager brides have outstripped the groom in racing to the waiting boat. The reluctant bride seems to be as rare as parents who are unreconciled after a few days to their daughter's marriage.

In this manner a man may begin to build his household with no necessary assets beyond a price of forgiveness that his confederate must negotiate with the bride's parents. When they approve the marriage, the price can be low, and a humble offering of flowers and incense plus 10 to 50 *baht* (current prices in the 1950s) forms the only nuptial ceremony. The couple then may join any household that can use its labor, possibly the parents of the bride or groom, or the household of some older brother or sister. With adequate food and accommodation for recompense, their help is welcomed in the fields and the kitchen. Then, as they contribute to the household of their hosts, they begin to accumulate the goods and tools for the day when they begin their own household.

Between wealthier households that may give property to inaugurate a household for the prospective couple, a price must be arranged in advance. Tradition prescribes a contribution of land from the bride's parents and a dwelling for the newly-married couple furnished by the groom's. Yet matching wealth is no prerequisite, for a rich landowner may welcome some poor but industrious son- or daughter-in-law to marriage in his house. Even in such a prearranged affair, consent of the bride poses a necessary condition for the match. Under the circumstances, divorce means withdrawing the separate properties of the principals and dividing any joint property that may have accumulated. Children follow the parent of their choosing.

From the foregoing the first principle of forming a household becomes apparent: All members join voluntarily. The husband and wife unite voluntarily and continue to remain together as long as they wish, for divorce is a unilateral decision by either party. Similarly other members of the household, such as Chaem's mother-in-law, may become members through mutual consent and remain together until one or the other expresses dissatisfaction. Frequently a younger brother or sister with spouse joins another household for a few years, and especially in the early days of Bang Chan they would have been a welcome addition. Even between parents and children a form of voluntary participation is symbolically maintained. Shortly after birth, the midwife, holding the newborn on a tray, formally asks, "Whose child is this?" The mother or some member of the household replies, "It is mine." Thereby a parent voluntarily expresses his consent to care for the child until it is grown or has reached an age to be farmed out to some other person. Then the child may be given to an aging grandmother who is lonely or needs a juvenile helper. This may last until the child's father finds it expedient to send him as a servant to the temple, where he serves some younger uncle or cousin in the monastic order by carrying the alms bowl on the morning rounds. Having provided some alternative care for a child, parents may leave him without remorse, and the children themselves, far from being slaves to a new master, may return to the parents when and if they are so inclined. Thus after the years of infancy, the parent-child relationship, rather than an inescapable obligation, becomes a matter of choice.

Siblingship is also voluntarily contracted. Though brothers and sisters are born to the same parents within a single household, a relationship within this circle is meaningless unless it implies more than propinquity. So children tend to form into pairs of adjacent younger and older siblings who play together, move off on little expeditions, and assist each other with chores. With other siblings much older or younger, they may cooperate a little for a short time or not at all.

Indeed, the Thai word for sibling, *phi-naung*, is a vague term of reference that applies to children of the same parents, half brothers and sisters, cousins in all degrees, their spouses, the siblings of their spouses, and the spouses of their siblings. From this point of reckoning, nearly every man and woman within an elastically defined age group is a potential sibling, yet the important spark lies not in this potential but in the fact that two persons bring themselves together as a voluntarily cooperating pair. Even when kinship cannot be traced between two persons who consolidate their efforts, they may address each other as siblings. So Chaem and his brothers, though misfortune may have thrown them together, agreed to cooperate on their journey to the wilderness. As the households of each grew in size and became more self-sufficient, the extent of mutual aid decreased, yet no one found it necessary to move apart from the other two.

A household thus gains members as its head forms paired liaisons with people who wish to come, eat, work, and live with him. The content of the paired liaison necessarily differs in each case according to whether a husband is dealing with his wife, his wife's mother, a child, or a younger sibling, for the particular circumstance that attracts and holds one will not attract and hold another. Some join already paired, so that a woman may move in, providing she can bring an aging parent with her. Thus Chaem may have persuaded his wife to move into the wilderness, as long as she could bring her mother. Antagonisms may also develop, as when a wife refused to allow a second woman to join except at the price of her own departure. However, all liaisons are polarized on a key person, who may be a woman with her liaisons to children and her siblings, rather than her husband. Then the household can often get along very well without this man; he would scarcely be missed.

A second feature of these paired liaisons is the automatic granting of authority of one member of the pair over the other. Among siblings the elder is in charge, and, except where old age modifies circumstances, a parent has authority over children. This authority rests on the assumption that one of the pair is in some sense a benefactor of the other, who becomes a recipient of some good. No matter how equal the effort of the two may appear, in last analysis one stands in authority. In the West we consider a reciprocal exchange possible only between cooperating equals; inequality of station seems to constrain us. The Thai, however, because they assume symbiosis to form the basis of reciprocity, deem an inequality to be essential. An aspect of this idea is expressed in the following Thai proverb:

> The earth is good because the grass protects it;
> The grass grows because the earth is good.

A rich man cannot help a rich man, but he can help a poor man; so a landed man can help a landless man, but not another landed man. The image of the good household reveals parents caring for their obedient children, older siblings tending their less competent juniors, and the able providing for the weak. To these various benefactors is assigned the authority to initiate action, while the recipient dutifully returns appropriate services within the limits of his resources.

On these kinds of reciprocities rests the substance of daily life. A parent feeds the infant in anticipation of having a future worker in the fields. An older sister passes a stick of candy to a younger brother in vague anticipation of a future occasion when he will rub her back, paddle her to school in the boat, or carry in a bamboo tube of water for boiling the rice. A tangible offering of some kind confirms the love that is returned in grateful services. Perhaps Western parallels occur in the respect of a young physician for an older colleague who offered kind advice; the eagerness of the Victorian Englishman to serve the Queen; or the devotion of the Napoleonic soldier to his Emperor. Phillips, in his study of Thai peasant personality, tapped some of these sentiments through the sentence completion technique, as shown in the following: "When he is in the presence of a superior, he feels . . ." Sixty-four per cent of the respondents completed this sentence by saying something like "pleased to see the face of a man who has had good fortune." Another 25 per cent felt frightened or anxious lest they be rebuked or blamed; the remainder felt unafraid, composed, and enjoyed his presence. Phillips also posed the sentence touching on the other side of the reciprocity: "The best way to treat a subordinate is . . ." Here Bang Chan people said: "We have to speak to him nicely. Agreeable words alone can make people like us (pp. 144–156)." The other touched on feeding a subordinate and not overworking him. So authority wears mainly a benign mien, and he who would go it alone resembles the person who would walk rather than ride a hundred miles.

A third facet of these paired liaisons, always implied and rarely voiced, rests on the understanding that the reciprocation may end whenever either party wishes. No obligation prolongs a liaison beyond its time; no moment is too soon to terminate a painful meeting. So, as we have seen, parents walk away from their children, and adolescent boys from their parents, while daughters unexpectedly elope in the night. When a liaison sours, no claims of flesh or blood are affronted, since kinship is but a tie of personal experiences shared while in the same household or the same womb. And despite these invitations to instability, households seem only slightly more fragile than elsewhere. As liaisons are recognized as fragile, they may not be taken for granted but need frequent reaffirmation. Moreover, a man takes pride in the number of his liaisons and the intensity of reciprocation. If they dissolve, the fault is usually his. Young couples break up rather frequently but much less often after second marriages, as if they were better prepared to prolong a union. Of course, not everyone is free to abandon a languishing relationship. Chaem's old mother-in-law may have had to put up with considerable abuse simply because she had no other haven. A poor cousin may work long at unpleasant tasks to reciprocate for food and clothing provided by a rich elder who would easily welcome a score of impecunious kinsmen to replace him.

Thus we have depicted the kinship system as a set of voluntary reciprocities between pairs of people, where authority is granted the provider of benefits and where the liaison may be terminated unilaterally by either party. Though often implied in the foregoing, we must state three important corollaries explicitly: (1) The greater the resources that a man has at his disposal, the greater the number of reciprocities that he can form. Thus men with abundant resources quickly find themselves surrounded by kinsmen and quasi kinsmen offering their services. Poor men with less to offer must wait and bargain to extend their numbers. (2) The greater the resources, the more enduring are the reciprocities that are established, assuming good stewardship. Poor households tend to lose members easily because the benefits they offer can be matched at many other places. Wealthy households need have little fear of competition; the less loved will be allowed to depart, while the more loved find their benefits increased. Perhaps the most important of these corollaries is this final one: (3) Household heads seek to gain and hold the maximum number of reciprocities possible within the limits of their circumstances. The system gives its greatest plaudits to those with the largest following. The poor manager fails to balance membership with resources, while the good manager gains and holds his members. But let him not be niggardly, for the man who fails to use his resources wholeheartedly for his followers may find himself as shunned as if he were bankrupt. With these rules is the game of living played; they differ from the rules, penalties, and rewards of the system to which we have grown accustomed, yet they too sustain a community.

What then does kinship mean? Certainly it does not specify the duties and expectations within a fixed group bound together as flesh and blood. According to the Thai, who draw on Buddhist tradition at this point, the body is mere nails, bone, hair, fluids, and so forth. Only the soul, which enters during the period of gestation, gives distinguishing characteristics. What is registered as amiable by two souls binds them together. Mutual experience begins with child and mother, first within the mother's body, then in the time of nursing and rearing. Growing up in the same household brings mutual experiences to the household members; so can attending school or serving in the same military company. When husband and wife feel particularly compatible, some say that they have shared experiences in a previous existence. Yet to fix and hold these possible connections requires some special expression of love, the giving of a gift and its reciprocation. The mother, giving her own food and body to a child, shows her love. So a sibling or a schoolmate with some smaller gift may also capture and hold the love of another. Then, depending on the relative age, sex, nature of the gift, and feeling of mutual confidence, the two, bound by amiable experience, are called parent and child, uncle and nephew, older and younger brother, or husband and wife. Kinship is psychic rather than physical.

The social forms produced by such personal liaisons cannot outlast the death of a member of a reciprocating pair, if they endure that long. Each person seeks to make the most enduring group he can from the persons within his reach. Once head of a household, a man finds considerable latitude of arrangement. Husbands may leave decisions to their wives, and even where a husband leads, the household is known locally by the names of both: Daeng and Chya, Bun and Weg, implying a partnership in marriage and control over the house. Widows often become managers of both house and field. A maiden aunt may attract a niece or nephew with spouse to come and live under her roof. The daughter of a rich landowner has married the industrious son of a poor man. However grateful this husband may feel to find a place for himself (and possibly his aging parents), his wife must learn to assert her authority gently to cause no loss of masculine face. Yet a decade later, when they have grown in ease of relationship, this former poor boy has become a household head.

Such flexible bonds between household members help rice growers meet the normal variations of fortune. When children are many in the household of a shifting cultivator, he can extend the area under cultivation and set each one to carrying home the grain in baskets proportioned to the child's size. Even transplanters with small, limited fields seize the opportunity of having nearly grown children by renting additional lands to work. Should his children subsequently move away, the owner need not contract his fields to a minimum, for he can muster an adolescent or two from some poor sibling's teeming household. At times of crop failure, household members may be released to go elsewhere, and when better times return, they may come back without embarrassment. All understand that householders seek to hold as many members as possible, so that departure, softened with regrets and some tangible parting gift, need not sever ties permanently. The departing ones recognize their peripheral position under a given roof, since the most dispensable person had to be chosen from among many, yet leaving for a better opportunity in another area also signifies a parent's concern for a child's well-being. A Bang Chan rice grower, telling of his childhood, upbraided his parents for refusing to give him into the care of a powerful government official who wished to "adopt" him. The storyteller commented, "My parents could not have loved me very much."

If households can add an extra hand or two each year and expand the area under cultivation, we might expect to find a countryside filled with enormous households working large tracts of land. Actually the average household stood just below six members in Bang Chan of 1953, the largest rarely exceeding a dozen. Areas under cultivation by a single household averaged 5.5 acres, and rarely exceeded about 18 acres. In subsistence agriculture, droughts, floods, and pests destroy sufficient crops to make a large household with few workers rather precarious, but even under a market economy where people raise crops far more ample than what they consume, other factors limit household size. Here we encounter local standards of living: Though rice be abundant, are the vegetables, meats, and spices to be eaten with it more or less than what one's neighbor serves? Are there sufficient tools, draft animals, and other equipment? Do the local workers feel demeaned to flood a field by hand rather than using a gasoline-powered pump? What about clothing, pocket money, days off to visit the neighboring market? Moreover, before a dozen people have settled long together, households tend to break into separate groups, each with its own hearth. Such division need not bar joint cultivation of a field as long as the participating households dip grain from the same rice bin and share abundant perquisites, but in fact only the abject long remain in such a position. The energetic young seek to acquire their own tools in order first to become tenant managers of independent households, rather than quasi servants. With good luck they will become land owners, albeit small ones, for land is costly, credit hard to arrange.

Though individual households persist as long as their geniality and resources continue, a cluster of households tends to be less enduring. Each supports primarily its own membership, but unless significant differences in wealth are present, the ties between households are weak. Chaem and his brothers had nearly equal resources, so that no one of them could offer a basis for symbiotic reciprocity. We have seen how, at the beginning, they doubtless assisted each other through the familiar exchange of labor in building houses and clearing fields, when each in turn acted as host for the day. But as the work force under each roof grew with children, each became more self-sufficient. The majority of household clusters in Bang Chan approximates this pattern of nearly self-sufficient neighbors. They may have arrived together, lived together in peace, and found no pressing reason to move apart. Living is a little more fun with sociability and occasional help.

Sometimes a tighter hamlet organization does appear where someone with more than average resources has moved in. He has provided not only for the members of his own sizable household but for other households as well, households that are too deficient in land or other resources to carry on for themselves. They accept the rich man's contribution and reciprocate by helping plant and harvest his crop or in other ways as needed. Today the benefactor gives them money when they are short, contributes to the marriage of their children, and arbitrates their quarrels. As in Chaem's hamlet, they too may be brothers and sisters, but include more distant cousins if resources suffice. Because it contains certain common or public features in its setting, such a hamlet resembles more the ones to which we are accustomed, having at least a common wharf for the boats, a place for pasturing live stock, perhaps a shelter where persons may gather on a hot afternoon. If Chaem's hamlet could have afforded a few of these features, it lacked the initiative to build and maintain them, for no one was in charge over all three households. No one but a benefactor can muster the labor.

THE KINSHIP COMMUNITY IN ITS ENVIRONMENT

For at least the past 1,000 years Southeast Asia, excepting North Vietnam, has been a great forest with occasional savannah-like areas, a landscape lightly sprinkled with villages, with an occasional walled city. Yet the grass and jungle-covered ruins of past capitals are strewn up and down the Irrawaddy, Chao Phraya, and Mekong Rivers. Local ruffians arose who could attract enough stout-hearted fellow villagers with promises of riches to attack a neighboring village and carry off loot. If fortune favored, followers increased so that in a few years they might attack a major capital. Riding homeward on caparisoned elephants, they led as captives the men, women, and children of the vanquished city to the prospering town of the victors. There these newcomers worked to dig the moats, raise the walls, construct temples and palaces, weave silken vestments, and on festive days dance before the Lord of the new city. Then, and in many places up until 1930 and even today, wealth was and is to be counted not in ubiquitous land but in manpower. Without these hundreds of hands gathered to a single spot, the building of these cities would have been impossible. So capital cities like Pagan in Burma, Angkor in Cambodia, Ayuthia in Thailand grew and fell into decay under the assault of some vigorous new leader. Few lasted more than a century or two, with Ayuthia's three centuries of repelling invaders a remarkable exception.

In such a land of scarce population, the gatherers of people became dominant, and though not everyone could build a city, many attracted and held enough followers to make living a little easier. The man with a single spouse and their eventual children could build his house and feed his family, but would enjoy few of life's amenities. The rough tamped earth in his hut compared poorly with the polished smoothness of sawn teak. Worn and faded cotton loincloths were all that a poor man's man power could occasionally wheedle from a poorly provisioned market; but His Excellency's silks came from his own establishment where retainers plucked mulberry leaves to feed the worms, spun the raw silk into thread, and then wove gowns and scarves. The poor man's wife served his rice and vegetables on a banana leaf, for he had no potter to make bowls and plates. The poor man's children were too occupied gathering vegetables in the backwaters to coop up, water, and feed the scrawny chickens that scavenged for food in the dooryard and rarely produced either egg or chick. There were neither smiths nor silver to fashion the little boxes for leaves, lime, and betel nut that were politely offered to His Excellency's guests at the end of a meal. On these vast differences of wealth and the symbiotic potentials that grew from them arose the fitful civilization of Southeast Asia. Yet the social system was open to all; each additional hand that could be lured to the poor man brought his household a little nearer achieving some slight portion of the greater amenities.

Certainly Chaem and his brothers were but slightly removed from the abysmal bottom. Few strangers passed their huts, and they drew on whomever they could. Slight differences appeared in these households over the years. Chaem with his wife and his mother nourished twelve children of whom eight survived to become adults. But most wandered away, as if his roof and garden could not contain them. Chaem's eldest brother had two wives, in addition to his aged mother. Though one wife was blind, she could make mats and baskets and bore one child. The other wife had three children, and they adopted another child. All remained in the neighborhood until their father's death. This eldest brother seems to have been the more genial person and energetic provider, a difference that Thais sometimes explain on the basis of position in the family. The eldest child becomes accustomed to self-sacrificing responsibility for others at an early age, for he must give to hold the affection of his juniors. The youngest child, they say, is the spoiled one, always the receiver without having to cherish another, only to be punished by his parents for disobedience to his elder. Such differences in skill to form and sustain reciprocating liaisons with others may well account for the dwindling of Chaem's household and the flourishing of his eldest brother's. And as for the middle brother's household, there were five children by a single wife; some children stayed, others wandered off; he held his group together but never expanded it.

Chapter 6

YEARS OF BROADCASTING: 1890–1935

In 1872, King Rama V (1868–1910) proclaimed a gradual manumission of slavery, whereby the child of a slave, specifically a child born of slaves during his reign, would become a freeman on reaching the age of 21. Enthusiasm for this measure ran high among the close associates of the king, and many, wishing to set an example, freed their slaves within a few years. About 1875, some 10 households of former slaves of the Viceroy set off for Bang Chan to enlarge their fields beyond what was possible in the old hamlet on the banks of Saen Saeb Canal. In the new area they set up a small clustered village that eventually reached about 30 households. In the circle of outlying land they raised their rice by shifting cultivation. Among the later arrivals was a freed Laotian slave from Bangkok who had received from his former master a plow and a water buffalo to help him make his way. Instantly this man became leader, since people wished to use his plow. In return for plowing a *rai* of land (0.40 acres), village members tended his rice plot and gave him a *thang* (22.2 lbs) or two of unhusked rice at harvest. Doubtless this benefactor was seated in the place of honor at local festivities, invited to arbitrate disputes, and asked for the loan of rice by households in distress.

The establishment of this village had little permanent effect on Bang Chan, for within a decade all but two or three of its houses had vanished. As soon as a man could buy a buffalo and a plow for the equivalent of a few short tons of rice, he set off on his own, followed by the households of siblings who wished to join him and use his plow.

Even though these people had the requisite tools and land to cultivate as broadcasters, they continued, like shifting cultivators, to drill holes in the plowed earth and drop in seeds. What they lacked was the man power to weed and harvest more than a *rai* or two of land, though the plow made twice that amount available. Later a Laotian household with its own plow and five or six strapping children might have turned to broadcasting, yet the first true broadcasters of which we are certain appeared as a relatively affluent group from Khlaung Tej toward the end of the 1880s.

THE KHLAUNG TEJ PEOPLE

While Chaem and his brothers had made their way to Bang Chan from Samsen, north of the king's palace, the people of Khlaung Tej came from an area on the Chao Phraya River a few miles south of the palace. The city was expanding, and prices of land rising. Title deeds were nonexistent; courts of law lacked jurisdiction, so that people defended their claims as best they could. Any occupant of land without a protector was in danger of being dispossessed. A man told of his maternal grandparents:

> My grandmother Nu sold her land but was cheated by an official who forced her to accept a price below its real worth. My grandfather Maut lost his share in his father's land when his brother gambled and lost. The garden, orchards, and rice fields all had to be sold.

For most people all this might have made little difference, since they could simply move to the broad wilderness and take up vacant land. These Khlaung Tej people, however, sought to forestall these unpleasant dislocations in the future by protecting their holdings more adroitly. Within the group were two or three men who served the king by mixing medicines for three months each year in the palace dispensary. The eldest was Mau An, whose title "mau" indicates that he cured sickness, perhaps by herbal medicines, esoteric formulas, or exorcisms. He led the delegation to petition help from the dispensary chief in acquiring and insuring occupancy of land in the wilderness. Since no

Bangkok official held the wilderness, the backing could only have been a verbal statement asserting that anyone wishing to dispossess this man would have to reckon with an official in the king's palace. Against all but the most desperate intruders, mention of the palace doubtless offered some security.

The palace official put the migrants in touch with a man who knew the way to good land, which, as it happened, lay on Kred Canal in Bang Chan. They soon returned with an appropriate gift to the official and gave a generous but less impressive tip to their guide. The descendants of these men report that their fathers and grandfathers thereby "bought" 50 *rai* (20 acres) of land, though many of their neighbors merely took over adjoining lands.

Unlike Chaem and his brothers, these new migrants were not in the least abject. The Bangkok order of affairs had rejected them, too, but on easier terms. These yeomen, cultivators of the urban periphery, had been bruised by receiving less compensation for quitting their lands than they deemed proper, yet theirs was not an ignominious retreat with little more than their boats and clothing. The newcomers brought as much as they could by boat, the backs of buffaloes, and the backs of men: plows, pots, mattocks, knives, mats, and baskets; some ducks, some chickens, their dogs; plantings of sugar cane, of betel nut. They were transplanted households on their way to fullness. They moved to settle more or less as a group, each on his own terrain, yet within easy calling distance and a minute or two apart by boat on the canal from a neighbor. Their bamboo and thatched houses, many of them stiffened with timber frames, were as large in the beginning as the later ones of Chaem and his brothers.

Let us make no mistake: these people faced a wilderness no less forbidding than did Chaem and his brothers. They too dug out the accumulated plant debris in the natural run-off channel to bring their boats to their house sites. Elephants still crashed through the brush in the great open beyond their houses, and snakes lingered quietly among shading reeds waiting for a frog to appear in the ooze. However, these were people from the outskirts of the city, accustomed not only to sustaining themselves from their fields and gardens, but also to considerable exchange with the markets. They required more frequent renewal of clothing than the single loincloth per year of the poor. Rather than grind a mattock down to a stump, they discarded the old for a pristine new one. They planted rice fields of 5 or more *rai* (2 acres) and worked them with the abandon of broadcasters to bring in 3 to 4 short tons of unhusked rice. So their storage bins were usually full, and they had ample rice to exchange, and some left to make into liquor for their own consumption.

Had one watched these people, with their slightly greater affluence, begin to clear away the brush, their work would have looked much like the shifting cultivators in Chaem's hamlet. Yet their larger, more stable households and better organized hamlet brought more hands to a collective job performed in turn for each household. After clearing, they too burned the dessicated grass and brush. Work, however, did not stop when 2 or 3 *rai* (0.8–1.2 acres) had been cleared, but continued on to patches twice that size or more. Unlike shifting cultivators, these people depended on the flooding for the success of their crops, so that each household head had previously sought the low-lying areas for center of his field. There the waters would rise earliest and linger longest to sustain the plants. After the burning, each household carried on by itself, and, instead of trying to prepare the soil with a hoe, hitched a team of buffaloes to the plow, and on another day to the harrow for smoothing the clodded earth. This done, the individual householder stepped out to broadcast carefully selected seed across the clearing, scattering handful after handful from a basket. A light harrowing then camouflaged seeds with enough dust that the predatory birds nesting in the nearby thickets passed over, thereby allowing seeds to sprout when the rains came.

Long part of the agricultural scene, these Khlaung Tej people were also more knowledgeable about ritual matters than Chaem and his brothers. No doubt, before coming, Mau An had copied (for he was literate) the requisite pages of the almanac found in most temples, which indicated the auspicious days to begin plowing and planting. This book would also tell him the proper direction for the first plowing, so as not to run against the scales of the great world serpent, the Naga of Hindu mythology, which rotates its position through the seasons. With decorous politeness he could ask the local guardian spirit of the earth for permission to clear the brush and stab the earth with his plow blade. He and his neighbors also knew how to summon the rains. A group of lusty young men, carrying a caged female cat, would sing as they went from house to house, inviting the residents to throw water on the caged animal and to give them a drink of rice liquor. The noise of the procession, with ever more boisterous singing and yowls by the angry cat, was intended to reach Nang Megtala, Goddess of Rain, who, wishing to quiet the uproar, would send heavy showers. Had she been unmoved by these means, they might have built with clay in the open fields a lewd pair of copulating figures, whose affrontery would spur the deities of the sky to wash them away. This failing, there remained the irresistible verbal formula to command the rain, which was shouted to the heavens by some knowledgeable one, but few availed themselves of this remedy, because the caller risked being struck by lightning.

Of course, the crop could never have reached maturity without the appropriate offering to the Rice Goddess, Mae Phosop; so on the days of harvest this hamlet had its special festivities. Such festivals were necessary not only to speed along mechanically repetitive cutting, threshing, and winnowing, but also because the larger fields of the broadcasters produced crops two or three times as large as those of shifting cultivators. Hands beyond the individual harvest were badly needed. Whichever household was ready to reap its grain ran up a flag in the misty dawn to tell the workers where to come. Should two or more households have been ready on the same day, workers went first to the field of the elder. Women and children followed to carry away the sheaves from the reapers and to spread them out on dry ground in a circular mass. Then boys came with their buffaloes to a stout post set up at the center of the extending circle of sheaves. Each boy tied his animal to a single rope that pivoted about the post. At a command the row of buffaloes began to move, trampling loose the grain from the sheaves. The outermost creature had to run while the inner ones moved at a slower pace. When one fell too far out of line with the others, the laggard was shifted toward the slower-moving center. After an hour or more of trampling and shifting of animals to straighten the line, a loincloth was given by the host as prize to the owner of the most enduring buffalo. As children moved off to bathe their parched animals in the nearby canal, adults joined to fork off the straw and fill baskets with the grain, which they carried to the dwelling of the host. In the evening, girls tossed trayfuls of rice into the air to winnow away straw and dust, while boys gathered for songs and courting.

This too was a kinship community, but not every household in the hamlet counted its neighbor as fictive or factual kinsman at the start. The majority had come from Khlaung Tej in two separate clusters of kinsmen, whose precise connections could no longer be established in 1957. We assume they were unrelated. The remaining two or three households in this hamlet came from other backgrounds independently and claimed their own land without special help.

Within each kinship group the household heads enjoyed such equality of resources that economic symbiosis reinforced the hierarchy of age only in part. Yet they respected the kinsmanly order of seniority with intermittent reinforcements: an elder brother loaned tools to his younger brother, a mother sent her daughter to aid the household of her younger brother because his wife had just given birth, a younger sister spent the day helping prepare food for the harvesters at the household of her older sister. All were clearly aware that at the pinnacle of this hierarchy of age sat Mau An, whose house was erected first, who was served first when he came visiting, to whom one could take disputes and expect a reasonable settlement. They knew too that when Mau An was serving his three months each year in the palace, his brother-in-law through his wife's younger sister stood next in the hierarchy.

The initial distance between nonkinsmen with nearly equal resources was tempered by the need for harvest workers. The exchange of labor among households gracefully allowed each to act as host-for-the-day to all the others. Soon they were greeting each other, out of respect for difference in age, as "older brother" or "mother's younger brother," and gradually trust replaced apprehension. This direction was strengthened on holidays like the New Year festival of spring when young people came respectfully to pour a tiny cup of water on the shoulder of an "older aunt and uncle" before dashing out of the house each with his bamboo tube of water in uproarious laughter, sousing each other from head to foot as they ran.

Not many harvests and New Year celebrations later these separate households were in fact kinsmen by marriage. Good neighbors find little difficulty in settling the exact contributions of each to their children who propose to wed. Of course, these alliances between local households can be made among kinsmen almost as easily as among nonkin, for here children of siblings encounter no firm incest barriers. Soon new households of son or daughter began to appear alongside those of the parents, and the yeomen concerned with making the wilderness habitable became a community of kinsmen in fact. By this time each owner held some 50 *rai* (20 acres), using at most 10 *rai* (4 acres) for cultivation, so plenty remained in brush to pasture the buffaloes and for future tenant households.

During the very period that this kinship community was solidifying, a new wave of migrants began to pass through the area of Bang Chan. Most were scouting land to claim, and finding only the less desirable land available in Bang Chan, moved from this new borderland to the frontiers farther to the east. The normal trickle of persons displaced from expanding Bangkok was being augmented by a new stream of liberated slaves. Impatient to sweep away the lagging remains of slavery, King Rama V about 1895 began taxing the masters who, short of cash in a changing economy, chose to send this body of dependents on its way. A few groups of migrants, having arrived too late to go farther and still plant a crop before the rains, were allowed to build a hut in some part of the unused land of Khlaung Kred, clear a *rai* or two to grow their food, and thus sustain themselves until the coming season when they might go on to unclaimed land. Doubtless many a landlord loaned rice to tide them through the growing season until it could be repaid after harvest. Most tenants moved on at that time, but some stayed, among them the dispossessed like Thiam, who recalled:

Father had 30 *rai* in Khlaung Tej and gave it as security for a loan which he could not repay. So we came to Bang Chan and rented land. I was 14 at the time. In our household lived six people, including my parents and two sisters, my brother and myself. There were deer here at that time, to the north, but we did not hunt very often. Each day we killed 9 or 10 snakes. To the south was heavy bush, so that we could not walk there. We cleared the land cooperatively. We planted rice with a stick in a seed bed and transplanted it after the rain had softened the field, because we had no buffalo. We used little land but got 40 to 50 *thang* per *rai* (1.3 short tons per acre). We paid rent of 10 *thang* per *rai* and sold rice to earn money.

We note in passing the continuing presence of the wilderness at no great distance and the transplanting technique that maximized the crop on a parcel of land shrunk to the minimum by rents of 25 per cent of the crop. On the basis of Thiam's estimated returns and our estimate of 1.1 lbs. of rice consumed per day by each person, four *rai* (1.6 acres) could sustain the household, pay the rent, and leave some small margin to trade for necessities.

Living as tenant was adequate but scarcely abundant. Should the crop fail, landlords were accustomed to diminishing or even forgiving the rent, yet for all householders the first duty was to provide for their own households and second, if tenants, to help the landlord. When the landlord's buffalo shed needed mending, the tenants sent a boy to lend a hand. They contributed to his harvest crew, but since their fields were small, they expected no help from him. He might draw upon them to help prepare food for a New Year's feast, to which they willingly went in order to enjoy a bounteous meal and at its conclusion bring leftovers back to the household members.

In hamlet-wide events they were less likely to participate. No member of a tenant household was snubbed at a winnowing party. And though no social barrier kept them from driving another's buffalo to thresh the crop in the gay exchange by members of the core community, a poor household had little time for more than keeping its own roof intact and serving its landlord.

A tenant occupies by dint of his fewer assets a lower position in the hamlet group than a landlord; yet a symbiotic relationship binds the two together. However, this fact of hierarchy implies no degradation, no obsequity, and little competitive envy, for everyone, from king to woodchopper in the forest, also has a superior and stands above another being. The ranks of gods and demons tower above man, below him those of animals from noble elephants to fleas, all in the great ordered hierarchy of existence. According to how well he performs the duty of his station, each is reborn in a coming existence. The buffalo that plows the field, submitting patiently to the whip, may be reborn a tenant farmer. Let the tenant farmer patiently pay his rents; in the next life he may be reborn a landlord or even an orchardist. The pains of human existence expiate the sins of past lives, and sometimes the expiation ends abruptly. In Khlaung Kred an industrious tenant's boy married a daughter of Mau An, and after a few years of service in the home of his wife, his father-in-law settled the young couple on land bought for the occasion. No one raised an eyebrow to imply that Mau An's daughter had married beneath her station, that Mau An had been excessively indulgent, or that the son-in-law was an adventurer. The achievement spoke for itself in showing that some of the son-in-law's sins had been expiated, and thus he could move to a better station.

A fall in station is quite as possible. A daughter of one of the founders told succinctly of neighbors who came with her father:

> Thim and Chum were poor, gambled and drank a lot. They pawned their land to my father as security for a loan, and then they could not repay. They moved away when they lost their land. Chum liked to live a sporting life and spend his money. He had fun as long as it lasted.

People speak not only of expiating sin through suffering but, when misfortune strikes, "running out of merit." These gamblers had fared well in proportion to their virtue; unhappily, merit did not suffice for a whole lifetime.

About this same period at the hamlet of Chaem and his brothers, who continued to live as in the past, a stream of migrants also passed, and some gratefully stopped to clear a rice patch in the tall grasses. Yet most of them moved away after a year or two, little lured to remain. No household in the hamlet had built up sufficient assets to embrace another in more than transient symbiosis. What could one expect from fishermen? Anyone can fish; it takes little capital, few assets. Moreover, in this hamlet quarrels broke out and smouldered. Not even Chaem's eldest brother with gray streaks in his hair could dampen the bickering between households. So the lowly hamlet continued little beyond a sustenance level, attracting newcomers at about the same rate as older comers departed. Thus it continued until the day when Chaem's nephew named Phlym was appointed commune headman in the newly organized district of Minburi, a tale to be told presently.

In comparison with this raw and edgy hamlet, the people of Khlaung Kred led their lives in great serenity. Both hamlets had a core of siblings, all with approximately equal assets. The difference between them lay first in Mau An who, though he could muster only a few more *baht* than any other person to loan to a needy neighbor, could also cure sickness. He performed a universally needed service that commanded respect. The guardian and tutelary spirits who helped him in his curing could be inferred to be formidable, since his reputation as curer was untarnished. They could protect him from danger. Moreover, working in the palace cultivated a self-assurance vis à vis his juniors that counseled compliance with his will.

In addition, the landlords with their tenants enjoyed greater wealth than the fishmongers of Chaem's hamlet. The latter bargained in the markets with no advantage and little scope for charity, haggling over each imperfection in weave and each inch in length every time they purchased cloth. With little enough in their houses to sustain themselves, they frowned on loaning things; thus they rarely commanded and often had to submit. Living in Khlaung Kred with its more ample margin was gentler, more yielding, more harmonious. It might be troublesome to postpone the harvest with one's own tenants, and join Mau An's harvest crowd, but if this pleased the old man, it was worth doing. Even here a household head could occasionally call on the tenant's boy to take his place among the reapers instead of gritting teeth throughout the day in the hot sun. For the sake of sociability one might gamble away a colored scarf or two and still be able to dress well enough to join in the next festivity.

Of course, this rustic idyll could not endure. In 1908, the water gates for a simple irrigation system built by the Thai government were complete. They held the flood water of the area for a few weeks longer in the fields and thus brought secure water levels to the higher croplands. But this added water proved an inconvenience at lower-lying fields where no dry land remained for joint harvests. Khlaung Kred people had to take the sheaves by boat to the only dry spot for threshing, the house mound of the owner. There fewer workers with a single team of buffaloes worked longer hours less gaily to thresh the crop. When Mau An died and left his lands to be divided among three daughters, no one person again achieved comparable influence. Each of the original migrants then settled into greater self-sufficiency on his own lands and lived surrounded by his tenants and children. With the demise of the older generation, these children inherited smaller acreages, so that they no longer needed the cooperative labor of the community. Some were profligate and had to sell their land to tighter-fisted siblings in order to pay debts. Then land size expanded a bit with fewer occupants, but still the younger householders were entangled in the management of their own fields. Diluted by newcomers who shared little experience with them, Khlaung Kred lost its form as a community and became small clusters of kinsmen in a sea of cultivators.

THE MONASTIC COMMUNITY

Risk and safety, together with their psychological counterparts fear and confidence, form dimensions of all social systems. Thousands of American city dwellers walk daily among swiftly moving automobiles with considerable confidence, though pedestrian deaths are a daily feature of the streets. Similarly, rice growers of Bang Chan wade through tangles of growing rice without special concern for lurking cobras whose bite kills someone every year in the general area. Not only are these fears and moments of confidence a function of a culture (many Americans would hesitate to walk through a rice field), but culture itself arises in part from these emotions. Our apprehensions over automobiles have produced massive legislation that bolsters the confidence of pedestrians, while in Bang Chan timid people rely on charms and tattoo marks to reinforce the courage needed to do their work.

We may transpose these observations to the styling of relations between people in a society. Americans are encouraged to feel safe by relying on themselves to attain their goals and to sense a risk when depending on others, but the Thai reverse the paradigm, feeling more secure when they have the backing of a strong and successful superior. Alone, they say, one is helpless, and we have already found an example where Mau An regained confidence by fortifying his land claim with the backing of a palace official. Chaem, on the other hand, had to live exposed to risks without a protector.

The search for powerful guardians to insure the success of one's undertakings is a theme of Thai social life and has its counterpart in a group of practices concerned with supernatural guardians. Almost anyone can invite a number of protective spirits, but a specialist's aid augments the likelihood of success. Requests tend to be more effective when addressed in elegant or archaic phrasing by one learned in these matters. Some awesome spirits can only be reached by esoteric formulas, but the spirits in charge of rain, crops, households, and certain amusements are fortunately quite accessible with the ordinary language.

To invite a spirit to guard the house permanently, a post may be set at the northeast corner of the dwelling and, on top of it, a tiny house, big enough for a bird. Next to this one sets a tray with incense, fish, pork, and liquor, for spirits consume the unseen essences that we can sometimes smell. Then, kneeling on the ground with flowers and incense sticks, one intones as politely as possible an invitation to the local guardian of the soil, Phra Phum, to come and reside in the newly erected little house. Some people report that after a few minutes a butterfly lights on the house, showing that the guardian has responded. Such a protector, it is said, keeps a household free of strife and theft by giving the would-be thief the illusion that the house is lighted and filled with people. Later, when a man sets off hunting game, a woman expects to give birth, or a person is seriously sick, an offering to Phra Phum will assist these affairs. Guardian spirits may appear in dreams demanding offerings and have been known to make people who affront them sick or mad.

To these supernaturals may be added the range of amulets empowered to protect through the force of words breathed into them that will foster kindness by strangers, honesty in business dealing, or immunity to armed attack. In this way households and their occupants have been helped to meet some of their many troubles.

A longer-term risk is ignorance of sin which prolongs and intensifies suffering. A sinful person may be reborn into greater than human suffering as a dog, a buffalo, or a leper. Men do well to learn that the cosmos is a place of perfect justice where laws of natural morality operate. The workings of justice are sometimes delayed, but the ruthless tyrant who lives on in his golden palace will most certainly receive his punishment by the time of his next incarnation. Man must be taught that good conduct leads to diminished suffering. An industrious cultivator, if very virtuous, may be reborn a civil official or, if unusually virtuous, a guardian spirit. This knowledge came from the great teacher, the Lord Buddha, whose immensely virtuous conduct in many lives achieved the ultimate reward of peace through non-existence.

Chaem and his brothers doubtless had their supernatural guardians and had learned to live with various anxieties without access to institutional remedy. The people of Khlaung Kred, accustomed to greater amenities than the wilderness afforded, sensed a void in having no temple nearby. They desired the chance to improve their position in the next life by making periodic offerings to the monks. Then very specific dangers might arise, such as burying a corpse without a Buddhist monk to say the mantra that forces the soul to depart. A temple provides this indispensable ritual and in addition offers facilities for cremating the body so that the wandering soul is literally dispossessed. The problem was genuine, for households and even whole hamlets have been panicked by the appearance of ghosts that failed to realize they were dead. As we shall see, temples dispel other anxieties too.

Bang Chan had no temple until a certain vendor of pots, hoes, rakes, and medicines, after many years of plying his wares from a boat on the canals, settled in the hamlet on Khlaung Kred. As with many aging people, his thoughts turned to the next existence, which he wished to make as comfortable as possible. This was the reason kings and ministers of state had built the glorious temples of Bangkok. So he too with his few savings resolved to build a modest temple in Bang Chan. To this end he purchased in the early 1890s about 15 *rai* (6 acres) of land for 12 *baht* (at that time about U.S. $6.00) from one of Chaem's nephews. Others interested in adding to their balance of good conduct by "making merit," as these pious acts are usually termed, lent their strength to mound up the land above the flood mark. On this mound the vendor and his fellow merit-makers constructed a simple congregating hall where a monk could reside in the shadow of the Lord Buddha's image. The building ready, the old vendor led a delegation of men to neighboring Wat Bam Phen to invite a monk to come occupy the new temple.

Having a temple attended by a monk, people felt more secure. They could make merit with their offerings and listen to sermons through the circuit of holy days: the April New Year, the Lord

Buddha's day of birth, enlightenment, and death in May, the Lenten retreat during the rains, and the November Water Festival. Perhaps this merit would influence the growing season, too, and the cultivators could gain further help for the crop through the monk's prayer for rain:

> The Lord Buddha is omniscient; the precious Dharma [doctrine] is the highest thing in the world; the holy Sangha consists of those established in the Way and the Fruit. These are the Three Gems. By virtue of the power of the Three Gems may rain fall in due season. May good fortune come to all!

Not only could the presence of the Three Gems bring the rains; it could also keep the water in the fields long enough to bring forth the grain, prevent buffaloes from dying, keep the household healthy enough to work in the fields, and spare one from disastrous storms and floods. From these Three Gems benign influences also suffuse generally into the world, so that husbands treat their wives more charitably, neighbors become more compassionate, and hamlets more considerate of other hamlets.

The third Gem, or Sangha, as the monastic order is called in the Pali language, reaches back in time to Gautama, the Lord Buddha, who established a mildly ascetic discipline as the way to understand the doctrine. An aspirant petitions a monk of standing to teach him the proper words for addressing the members of a local chapter: "I come to the Blessed One who long ago attained Nirvana and to the Dharma and to the Sangha in search of refuge. May I, O learned one, enter the sacred order of the Blessed One?" He learns the formal replies to the questions he will be asked: whether he is a human being (rather than a demon in disguise), free of bondage, and unencumbered by debt. He vows among other things to fast from noon until the dawn of the following day, to wear only the yellow robes symbolizing the shroud of a corpse, and to avoid contact with the opposite sex. On the day of formal entrance into the order, the suppliant arrives in procession with kinsmen and friends to bear his robes, begging bowl, mattress, pillow, fan, and the other accouterments of his future position. In stepping across the threshold of the sanctuary, he leaves behind these friends and kinsmen to join a new community. There in the cool, semidarkened room the assembled monks sit cross-legged on the platform before the images of the Lord Buddha. In this quiet hush the suppliant states his petition, is clothed by attending monks, and repeats his vows.

During his stay at the temple a monks learns the prayers to be said in the darkness before dawn, to chant blessings for the dead, to walk with lowered eyes on his morning rounds for alms, and to follow the "Way" circumscribed by the 229 rules of the order, which have come down almost intact from the founding of Buddhism some 2,500 years ago. Infringements of the monastic rules are daily confessed to a committee that disciplines infractions.

In Bang Chan of the 1950s we were surprised to find the life of a temple not gently austere but easygoing, aside from celibacy and restrictions on eating. Monks seldom spent their afternoons at study or meditation. Some turned to repairing boats, some set off to visit the nearby market center, others lounged and napped. All the while roofs leaked, half-hinged shutters rattled in the winds, walks overflowed with weeds, and no one moved to repair them. Young men who had left the order confirmed that they had spent much of their time idle, and Bang Chan seemed to be no special exception among temples.

Long accustomed to seeing the revealed truth propagated with Dominican fervor, we occidentals are little prepared for indulgence in these proportions, especially in a monastic order. The Thai are more conciliatory: The Lord Buddha is a teacher who in his compassion addresses the apathetic and bewildered monks as well as the eager and perceptive. Though centuries ago he established the monastic order for learning, the will of each monk to learn precedes understanding. No one can tell in advance; sons of rustic hamlets may be better prepared to receive the Doctrine than sons of talented *gurus*. So everyone follows the rightful path as far as he can.

A conscientious abbot seeks to hold regular chapter meetings to reinforce the precepts and to present each monk a daily opportunity for confessing his lapses. He further offers the monks a chance to listen to sermons, to attend funerals, and thus to learn that self is an illusion. The rest, however, lies with the individual. As long as he does not flagrantly abuse the rules of the order such as those prohibiting theft and sexual gratification, a monk may remain as lethargic or energetic as his individual Karma determines. Some abbots urge their monks to help maintain the buildings; an energetic one even promotes the giving of additional buildings by laymen. Others, equal in piety, deem that this work contributes but little to teaching of the doctrine. If no one regrets seeing a temple collapse, the day is not far ahead when the monks will be hungry for want of offerings. If so, the heart necessary for a Buddhist temple has disappeared, and as an invited resident rather than its custodian, the abbot and his followers must seek greater receptivity elsewhere.

The abbot is head of the monastic household, where, like ordinary households, his authority extends as far as his reciprocities. Ordinarily senior to the other monks in years of service or knowledge of doctrine, he heads a hierarchy of religious proficiency. As with reciprocities anywhere, participation depends on continuing satisfaction. Senior or junior monks may terminate the relationship at will, though the junior continues to pay token respect or more to any teacher who has given him knowledge. At any time a monk may seek another teacher, move to another temple, or even formally rescind his vows before the local abbot, leaving the order.

Reciprocities within the order may extend beyond a local temple, so that the abbot of one is senior to the abbot of another, who in turn is the junior to a third. Thus chapters in temples scattered across the countryside form into hierarchies that reach toward a pinnacle commanded by the national patriarch, head of the monastic order. Monks intent on climbing toward these heights do well to move from Bang Chan to more important temples, become disciples of monks with standing, pass formal examinations in the doctrine, and acquire a reputation for skilled chanting of sermons, sensitivity in clairvoyance, or some other prized art.

This towering monastic community depends for its existence on satisfying the desire for knowledge of doctrine and offering people opportunities for doing recognized deeds of merit. In this sense it is an epiphenomenon, a shadow, of social substance, for temples and monks exist only in responses to these wants. A temple may own no property, and monks may not provide their own food. When offerings cease, no monastic storerooms are opened to feed the chapter. A monk depends primarily upon his kinsmen, and in Bang Chan parents eagerly fill their son's bowl with rice and curries when he passes on his morning rounds to gather alms. Through his months at the temple they also keep him supplied with cigarettes and tea, pillows and mosquito nets. Only households producing more than their basic requirements can afford to part with a productive worker for a growing season or longer, let alone sustain him through unproductivity. Chaem's hamlet, in need of every man and having little to spare, could ill afford this drain.

On leaving the order, each monk takes with him what he wishes, divides the remainder among his colleagues, and departs with no residue to be disposed of. Were it not for the continuous replenishment of its buildings and furnishings by merit-makers, the temple itself would crumble. Temple land has no owner. After clearing away the rubble of a vacated temple, anyone may plant there a garden or orchard. Only the sanctuary's hallowed site may never be used for a dwelling. No contracts bind the abbot to oversee the monks or to provide services for a congregation, nor does his resignation oblige anyone to seek a successor. Without corporate existence nothing remains awaiting a gaveled decision by some authorized body to dispose of its assets. The monastic community gains its continuity through overlapping, like the fibers and strands of a rope. Set this rope afire, and no ashes remain for the reliquary.

In Bang Chan's temple the first monk was followed by a succession of local abbots, among them a man who saw his duty to embellish the temple. He solicited contributions in rice as well as labor, and then added a sanctuary, where monks might meet and take vows for entering the order. The resulting porcelain-studded facade and niche with a bronze standing Buddha, whose hands were fixed in the gesture of stopping the charge of frenzied elephants, so pleased the people that they completed a reliquary in similar style on the following year. Though skilled labor was hired from Bangkok for these buildings, local people often helped. On their own initiatives they also built sheltering rest-pavilions for travelers along the canal and memorial pagodas to contain the ashes of deceased parents. With the simpler and less demanding acts of merit, Chaem's hamlet also contributed during the slack days of the dry season. However, it took a Khlaung Tej resident, employed for many years by the king for decorating royal temples, to paint a scene in the new sanctuary showing paradise with gods and angels living in peace among contented people, juxtaposed to hell where demons eternally tore entrails from hapless victims. Over the years the chapter of monks also increased from the original monk served by two temple boys, to as many as 45 plus somewhat fewer temple boys, housed in the 1950s.

THE GOVERNMENT COMMUNITY

Government, as a centralized agency supported by the population at large for carrying out public services, does not occur in this part of the world. Persons exist whom we may call ministers, governors, or officials of one kind or another, often able and honest men, but the services they perform reach only some of the people some of the time, while their revenues tend to come from the less aggressive or even defenseless. Ordinary people seek to proceed on their way independently, preferring to avoid officialdom, knowing that any meeting is apt to require dipping into one's purse, sometimes to recompense genuine services, often to avert seizure of person or property. Not even a minor official visits a village or hamlet without expectation of some return from the inhabitants, and inhabitants of most visited spots quickly give such a man the minimum needed to speed him on his way. Even if he comes with gifts of food, clothing, and livestock to distribute, the local inhabitants are stonily anxious to determine the obligations they are incurring. All assume, as if they were bargaining in the marketplace, that a service rendered requires some kind of return. If one wishes a judge to hear his plaint, one offers him a gift, not to bribe but to activate the interest of the court. A man wishing the police to apprehend a thief does well to gain in like manner the ear of the captain. Those who, in the nineteenth century, had to make annual payment of taxes or work off obligations to local officials by digging a canal or repairing a road, submitted to avoid some greater evil. Little wonder that many people preferred to avoid contact and take their chances against bandits or foreign invaders in the jungled backwoods beyond the scope of government.

Just as the monastic community holds a monopoly on the knowledge of making merit, so the government community holds, often precariously, a monopoly on the force of arms. In Southeast Asia, as we have seen, the founder of a dynasty was a successful general who built or captured a walled town to sustain himself, his men, and their wives and children. Each household within government reach was constrained to supply produce and labor for the town. Within this town the ruler, whether founding general or his successor, depended on a corps of subordinates, whom he rewarded for faithful services by grants of land, but more important in those days, by gifts of captives. In turn these subordinates, as lieutenants in charge of a quarter or ward of the town, supplied produce and manpower. With these subordinates the ruler formed the familiar relationship of symbiotic reciprocity. The same conditions applied as elsewhere: the relationship was voluntary and personal, terminable at will by either party. Little, beyond a ruler's vigilance, sustained these liaisons.

For example, in 1782, General Chakri raised a successful revolt against his aging and possibly senile superior, the King of Thonburi, to become founder of the present dynasty in Thailand. Of course, a ruler used every available device, terror as well as endearment, to surround himself with faithful dependents, yet not infrequently the man who almost succeeded in some lesser plot was cherished for his courage, punished, forgiven, and returned to office. The annals of each reign are dotted with instances of this kind, for, as in any reciprocity, devotion insufficiently compensated is valueless. Government was (and is) the followers of a war-lord type, a brittle community, guarded and suspicious, eager to seize opportunities to expand control, careless of suffering outside the group, and occasionally indifferent within.

The nineteenth century European advisers to Thailand, who counseled with images in their heads of hero kings ruling enraptured subjects, deemed the governing "ineffective and backward." They recommended centralizing authority through a civil service that would eventually reach the remotest hamlet in the hinterland. They advocated a police system, judicial reform, public education, railroads, an impartial customs, and a host of other services. These additions to government would be paid for by taxes collected from all people and doled out in budgeted amounts by a central treasury. In 1892, Rama V appointed his half brother Prince Damrong to form the first centralized ministry of internal administration, which gradually replaced the quasi-hereditary governors of the outer provinces with civil officials responsible to the crown.

No single system of taxation, no central budget, no treasury developed, for no ruler could impose such a system on his rivals for power. Here we may skip past the particulars to observe that the audience of officials contained some who recognized the possibilities of establishing new monopolies, of new enterprises to support their followers, and of gaining access to money, which was then proving an effective agent for gathering manpower. By 1898, a local district officer had been appointed to build and occupy new offices and a residence in Minburi near Bang Chan. Within a short time he had designated hamlet headmen in his territory, and among the new commune headmen was Phlym, son of Chaem's oldest brother.

In return, Phlym became collector of land taxes among the six or eight little hamlets that had sprung up in the Bang Chan area. As representative in the locality for the district officer, his never-well-specified duties led him into crime suppression, assistance with land records, and census taking. Yet services for his superior also included sending men to prepare food at a district celebration, planting the district officer's vegetable garden, and acting as host for the district officer with his entourage on tours of inspection.

Statutes authorized Phlym to keep 5 per cent of the taxes, but they did not specify how much he should collect in the first place. Thus he held the power to ask a little more money from one hamlet and a little less from another. Besides, it was possible to overlook reporting to his superior a household or two that recently moved in and share some fraction of his take as a generous gift to the local headman. In bad years Phlym made friends by diminishing taxes, and in ordinary years by inviting the people to a celebration: perhaps a tonsure ceremony for his adolescent daughter, or the rites honoring his teacher of the occult arts. Food and drink were always abundant, sometimes embellished with a traveling theatrical troupe.

Within his kinship community, news spread rapidly of Phlym's appointment. Strangers soon appeared from afar, announcing themselves as distant siblings, in hopes of finding a place to settle. Phlym himself began to buy up parcels of land to rent and sell to these newcomers. Persons in debt came offering their services, like slaves of old, if he would but assume their obligations. Some this new commune headman turned away, but those who remained profited him with labor, rents, and an increased return from tax collection, as well as themselves by association with a man of rising power. From the original three households of Chaem and his brothers, the hamlet swelled to 30 or more. Phlym became arbitrator among disputants, banker and philanthropist for the indigent, defender, rebuker, and fixer for kinsmen in trouble outside his hamlet.

We are now in a position to appreciate more fully the position of Mau An and his neighbors in Khlaung Kred when they sought help from the palace to secure their land. Until the twentieth century in Thailand, no one owned land except the king. When a man paid his standard land tax of one *baht* per *rai* (roughly 50 cents per *rai* in those days, or $1.25 per acre), he paid a quasi rent for the right to occupy and cultivate a certain plot for one year. Settlers in remote areas without commune headmen paid nothing, for no one bothered to collect. Thus as long as population remained low, residents of the rural countryside encountered little difficulty in occupying a site indefinitely. Anyone with heart set on living in a particular spot might offer the occupant some compensation for the trouble of vacating it. The people who originally dispossessed Mau An in Khlaung Tej drove a hard but not unusual bargain in buying the right to occupy his old lands.

As soon as many people begin to act in this way, occupancy becomes unthinkable. The rules of the jungle are operative—desperate characters appear, their eyes on some already cleared field, their intentions backed by guns and swords. Then arises the need for champions whose might can stabilize the claims. People live in time. They plant their crops in expectation of harvest, build flimsy houses for at least a year, and clear the brush from the dooryard as if this act would last a decade. Without such expectations, which war and civil unrest also destroy, there can be no settlement. The justice of the stabilizing champion is less important. The more obstinate among the dispossessed may appeal to the local commune headman, and if his interests in the newest occupant have not already been awakened with suitable gifts, he may come to the rescue of the older one. Mau An, with backing in the palace, had plausible defenses in this game of "King of the Hill" played against a threatening new occupant, together with the possible collusion of the commune headman. Whoever the champion may be, and whichever side he may take, this backing, once given, has to be continued by continuing services.

The Department of Lands had sought a succession of remedies in the face of angry petitioners. At first a tax receipt was ordered as evidence of occupation, but then both contestants found they could produce them. When the department required a description of the land to be written on the tax receipt, two descriptions applying to the same land afforded no better resolution. Well-paid testimony by a commune headman quickly ended the litigation. Finally, in the early 1900s, an Englishman inaugurated the Anglo-Saxon remedy, mapping land claims and issuing titles, the system that had long kept the British Isles free of these disputes. When surveyors reached Minburi, government clerks grew fatter for a few years on fees for the titles they issued, and the new generation of cultivators plowed with peace of mind. Subsequently lands became divided among heirs, were pawned to some creditor, or merely sold without reference to the mildewing papers at the District Office. Then people began to realize that the peace on the land had not been brought about by the unusable records at the district office, but by paying the commune headman through these many years for the protection he could offer.

No less important for rice cultivation were certain extensions of the water-control system. During the 1890s, King Rama V granted rights to a Thai-German firm for a vast tract of land in the central plains north of Bang Chan. By digging canals in this waste, the area could be brought under cultivation, helping to settle many people. A considerable tract was presented to His Majesty, and persons in the know scrambled to buy up the remainder. On the day of completion a procession of monks, princes, and ambassadors in barges rowed by liveried oarsmen moved along the canals behind the king. Within a few months the lands were filled with tenants of the favored few, but two years later these tenants were moving away. Because the land was too high, the water level had not risen sufficiently to flood the fields for more than a few weeks. Engineers from the Irrigation Department were called to save the undertaking. The completion in 1908 of three widely separated water gates provided the remedy. As the summer floods mounted, the gates were opened, then closed at the peak to hold the water level at the requisite depth for the crop. After a lean decade, landlords were delighted with the return of their rentals.

These deeper waters also covered Bang Chan, leaving the fields knee-deep when the rice was ready to harvest. There, as we have seen in the hamlet on Khlaung Kred, they spelled the end of hamlet-wide harvesting. The last threshing with 10 buffaloes in line trampling out the sheaves heralded the demise of kinship communities in Bang Chan.

The government community served the kinship community by stabilizing the land claims of many people, doing so in the bittersweet manner that characterized most of its activities. Its services, whether raising the water level in the rice fields or drafting the ne'er-do-well into the army, were no less astringent. Only by joining the government community itself, as policeman, clerk, or commune headman, did one savor greater sweetness.

When embraced by the government community, rather than remaining its prey, lives of people in hamlets were transformed. Chaem lived long enough to see his few remaining children, nephews, and nieces abandon fishing and raising rice in the fashion of shifting cultivators. Phlym, the new commune headman, quickly acquired a pair of buffaloes and set his new clients to plowing up broad fields for broadcast rice. Indeed, he could not well have maintained a position of authority in his own hamlet, had he not permitted his siblings and cousins the use of these animals for their plowing. They too found their households increasing with one or more impoverished kinsmen, who preferred to place their lot in the fringe of the great man's already well-staffed household, rather than prolong their search elsewhere for security. Thus the hamlet of fishermen, in the fanfare of their good fortune, became another hamlet of broadcasters.

In affluence and influence Phlym's hamlet quickly overshadowed that of Mau An. Though the latter's nephews and nieces continued to spend three months a year in the distant palace, their wages and perquisites sufficed to bind only a few of the households together. Moreover, compounding medicines or fireworks in the palace gave them no easy access to the ear of an aging king. Palace protection over the lands of its servants may have worked well enough in the past, but in these newer days local administration had become the arbitrator. The new commune headman, welcomed at any hour by local officials, his hand on land taxes for all the hamlets in his jurisdiction, Mau An's among them, could easily embarrass, if he so desired, the king's faithful servitors through some trumped-up contest over delinquent taxes. As for local éclat there was no rivalry. Mau An might stage a hamlet celebration for sending a grandchild to become a monk at the temple, but commune headman Phlym fed many hamlets at his festivities.

SOCIETAL ADJUSTMENTS TO THE ENVIRONMENT

With change from shifting agriculture to broadcasting comes a heavier weight of population on the land. When man has driven away the wild animals and uprooted the wild plants that once helped sustain him, he becomes more dependent on his crop. Larger crews of workers must be assembled to work larger areas with bulkier harvests. The pattern of energy input requires greater precision. Confined by concepts of property and trespass, these broadcasters must live more tidily. We find a more stable, more enduring, and more controlled organization of cultivators and ask what made this adaptation possible in the easily assembled and disassembled kinship community.

First we must look to the intensity of reciprocity between a household head and the members of his household. What serves to strengthen the tie between husband and wife, between parent and child, master and servant, or landlord and tenant?

Within a household, the liaisons of father or mother to each child are personal and unique, the character of reciprocities varying according to the case. As one child may receive more attention than another, so a landlord may rent to one tenant at a different price and with different perquisites than to another. The Thai see no virtue in equality, explaining that since one child is more helpful in the household than the others, or one tenant more industrious, these virtues should be encouraged.

Each group contains gradients of attention and esteem, of fat and lean, of used and neglected reciprocities, whether the liaison is favored or tolerated. Then when we ask which are the more enduring liaisons, which are more lightly loosened in the face of some necessity, the answer is clear. The more intense reciprocities will remain, while the thinner and more distant ones are allowed to lapse.

We thus arrive at a rough generalization: The greater the wealth of a group, the stronger and more secure the liaisons of members to its head. As compared with the poor man, a wealthy parent gives more to his child, is likely to receive richer services in return, can exert greater controlling influence upon him, and can hold him for longer periods. Similarly, the tenant of the rich landlord can expect more frequent loans of a buffalo and more generous reduction in rent on poor years. Because of the manifest rewards for overt generosity that lures people to one's side, the ambitious man misses no opportunity for dramatic ostentation. Food and drink must ever flow richly, and he who would stint for a rainy day declares himself a second-rater. The risks in this social game hinge upon the soundness of a leader's judgment in taking the course that will sustain his level of ostentation. On his part the follower must distinguish braggarts from the bold.

The size, complexity, and strength of a group is not solely a function of the absolute wealth of the leadership. Context is also important. The valley where all households are rich will be socially less ordered than the valley with one rich household standing among the poor. The village of Laotians grew on the small difference in wealth between a man preparing his field with a mattock and the man in possession of buffalo and plow. As soon as someone else could own a buffalo and plow, he departed the hamlet.

Mau An and the members of his hamlet attracted more people for longer periods than Chaem and his brothers, because at first these people on Khlaung Kred stood high above the level of a squatter in the wilderness, but when other hamlets with wealth equal to Mau An's appeared in Bang Chan, a young married couple found no striking economic advantage to remaining at the old stand. In his turn, the new commune headman, having formed a connection to the coffers of the government community, polarized the area in a new direction. By implication,. social stability in Thailand and perhaps in the Orient as a whole may be better served by encouraging the development of differences in wealth rather than by striving for social equality. Societal order depends on a liaison of respect between haves and have-nots that contains only the slightest traces of humility and condescension. Those who see progress only in the rise of an egalitarian middle class may well be undoing the very symbiosis on which the order rests.

A household can further be stabilized through affiliation with some more prospering group. One need but compare Chaem and his brothers, living isolated in their hamlet, with Mau An's kinsmen enjoying connections to the king's palace. Likewise Phlym as commune headman could draw on the District Office better to defend his land claims, to avoid paying extra assessments by local officials, and to make robbery more hazardous for the robbers. The strength of the mighty thereby filters downward through the hierarchy to hearten the humble at its base.

The monastic community also strengthens groups by broadening the basis of reciprocation. To self-centered definition of advantage, it adds love and compassion; the return for good deeds need not be weighed only for this moment but extends far into the future. Buddhist faith sweetens the joining of people, and Thai declare, "Those who make merit together will meet in another life." Learned monks have noted the correspondence of household stability with the law of Karma. Birth and death, they explain, are painful. The more meritful elephants with a life span of 90 years suffer less than the less meritful gnats, painfully born to live and die in a few hours, perhaps then only to be reborn a gnat. So too household heads of greater merit (and hence of greater wealth and ease of living) outlast those of lesser merit. Through these longer years, their households too endure. On a clay amulet the following verse was inscribed:

> He who has love is beloved by mankind and angels;
> He who has compassion is cherished by parents and rulers;
> Therefore those with hearts filled with love and compassion
> Are embraced by the benign and will live long lives.

Nai Sud, a retired rice grower.

Dwellings stand among trees, face the canals, and are most easily reached by boat. In the center background a spire rises from the local temple. The bamboo house has had its roof renewed.

This old man owns no house or land; he finds shelter for himself and his grandson at the temple.

While rice sprouts grow in the seed bed the fields must be flooded and plowed for transplanting.

A rice grower addresses the guardian spirits before transplanting the sprouts into the flooded field.

When the harvested sheaves are brought to the threshing floor, buffalo trample the grain loose.

Workers lift the straw from the grain and pile it into stacks for the buffalo to eat.

Grain poured through this hand-cranked blower is freed from chaff and dust.

In her best attire, this woman ceremonially brings the gleanings from the field, for they contain the "spirit" of the rice. They will be tenderly placed with the seed rice to await the next planting season.

The market and rice mills along the canal in Minburi.

From Minburi the milled rice moves in barges along the broad Saen Saeb canal en route to Bangkok.

At the congregating hall of the Buddhist is the clearest representation of the cosmic serpent or naga. From the point of his tail above the roof peak, he slithers down the edge of the roof while his head rises at the eaves.

Corpses are often held in the morgue of the temple until after harvest when public ceremonies begin. At a cremation ceremony the bier is placed on the cremation platform. After mourners have presented final gifts in the name of the deceased, the decor is removed and the fire is set.

Young men aspiring to become Buddhist monks are feasted, shaved, and bathed in a day-long ceremony staged at a kinsman's house.

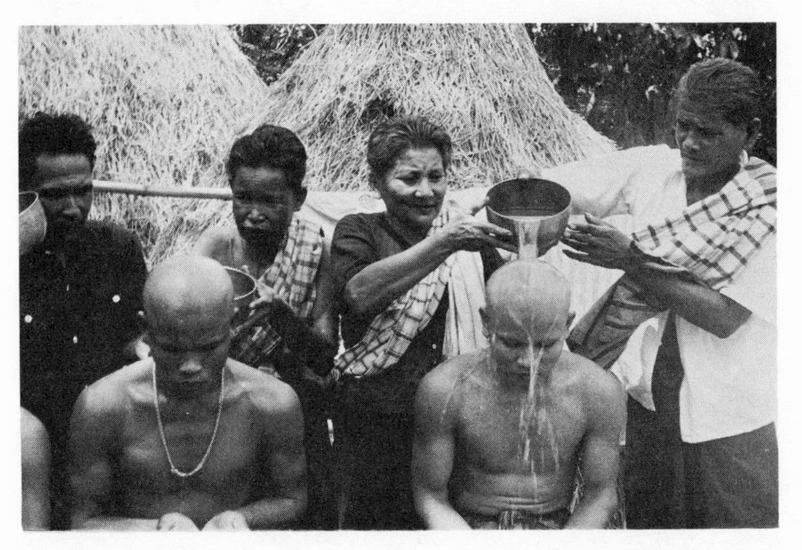

Musicians lead the gay procession conducting the candidates for the monastic order to the temple ground. When the rains begin again the procession moves in boats.

One may solicit protection from supernatural beings. Here pigs' heads and liquor are offered to sustain the protection of a guardian spirit.

Chapter 7

YEARS OF TRANSPLANTING: 1935–1970

Two steps were needed to transform Bang Chan broadcasters into transplanters. The first occurred in 1918 when the Armistice ended fighting in World War I. Then the flour mills and bakeries sought from overseas the grain to feed the hungering millions of central Europe. Dealers the world over, from Melbourne to Montreal, had long awaited the moment when the British naval blockade would end and mine sweepers clear a safe channel to the idle docks. Any old hulk that still responded to its helm was pressed into service. Any bag of grain in any dusty godown, clean or not too weevily, was in demand; agents offered unbelievable prices.

The news struck as the first rice barges of the year began sailing the new crop down the rivers and canals to Bangkok. Elderly Chinese merchants were suddenly cabled dizzying offers for hundreds of tons of rice from long-silent cousins in Hong Kong and Singapore. Their sons and nephews were soon fanning out among the backwaters even to acquire rice still green on the inflorescence. Bang Chan prices rose 8-fold, from 20 *baht* per *khwian* of the preceding year to 160 *baht* ($6.75 to $54 per short ton), and growers crowed, "We're going to be rich!"

Anyone with rice to sell during that first year counted his pile of silver *baht*, quickly converted it for safekeeping into a boat or a buffalo, and congratulated himself. But many a tenant's joy collapsed on hearing that landlords no longer wanted to rent them land. It stood to reason; when a landlord rented for a quarter of the crop, he could earn four times as much by working the land himself. So within a year two hamlets like Khlaung Kred lost half their households. House mounds of the tenants were leveled, bushy patches cleared, and pastures converted to tillage. Every scrap of land came under the plow, as hamlets thinned out, giving way to single dwellings standing alone.

To move the mountains of unhusked rice from the field to a Bangkok mill imposed extra work on the sails and rowers of every sampan and barge. Mills ran for months unable to polish and sack the mounting piles of waiting grain. Captains of anchored ships berated their harried agents ashore for unwarranted delays in loading. Boatyards were swamped with orders for new barges and tenders. Carpenters, masons, and mechanics worked from dawn into evening darkness to build new mills near the fields where rice grew. In Minburi they worked rapidly, so that before the final months of 1920 a new mill was ready to receive the crop from Bang Chan and its neighbors.

During the initial settlement of Khlaung Kred, households claimed about 50 *rai* of land, but a generation later the average holding was reduced to about 30 *rai*. The older generation could well divide the land into equal parts among all the children, as custom dictated, and certainly until after World War I all could have sustained themselves on such a holding. The postwar appetite for larger acreages and money to spend, however, made a mere 10 *rai* (4 acres) an insignificant holding. Many sold these small claims to some sibling and took off to the thinly populated periphery where cheap, uncleared land was still available.

We have seen in Table 4.9 that 35 *rai*, or 14 acres, required 292 man-days to grow a single crop by the broadcast method. Bang Chan informants speak of three able-bodied men handling a slightly larger field. Of course, the harvesting of 11 to 16 short tons of grain drew on the labor exchange between neighbors, but for the major portion of the crop season a man, his wife, and a grown son could manage fields of this size. If we include two more persons for the household duties and a child or two to tend the buffalo, then a middle-aged couple with one married child, whose spouse and child also shared the work, could barely fill the labor requirements. In tending these larger acreages, households thus grew from slightly extended to considerably extended families. Parents were pleased to have two married children, their spouses, and children to lighten the work.

Of course, these enlarged households were earning more than anyone dreamed possible. Soon staunch teak-planked houses with carved wooden nagas descending along the gables began to replace the simpler dwellings of timber and thatch. After harvest, peddlers, their boats precariously loaded with chests of drawers and storage cabinets, plied the canals making easy sales. Open-handed customers from the rice fields appeared at familiar stalls in the Bangkok markets to buy unexpected quantities of cloth. Holiday processions of merit-makers in handsome new boats set off with offerings to the temple, brightly colored scarves fluttering from the shoulders of the paddlers.

After 10 such years, about 1930, came the second step, a somber one. Then rice could scarcely be moved from field to mill or from mill to the harbor. With the surfeit of rice, prices dropped in stages down to pre-World War I levels—even to 20 *baht* per *khwian* ($6.75 per short ton). Empty barges lined the shores. On wharfs, sacks of polished rice stood unwanted among equally unwanted bales of latex and stacks of tin ingots. No one starved, for in the fields was rice and in the canals, fish; idle clerks had time to garden and raise pigs. Of course, the scarves on the shoulders of the paddlers faded in brightness, and travelers carefully saved their shoes by carrying them under their arms.

Cash, the addicting stimulant of social systems, had dwindled to a trickle. Households on the land harbored their young people with no place to go, yet return to subsistence was difficult. Many paid for their inability to curtail spending by losing land to their creditors. If becoming tenant on the old plot failed to subdue appetites for market goods, they lost their buffalo, and landlords brought in thriftier tenants. Even among the resolute, this thirst was only repressed, never obliterated. If prices were low, they reasoned, cash returns could be increased by growing more. This response was simple and direct, uncomplicated by worries over glutted markets or the economics of supply and demand. The moment had arrived to change from broadcasting to transplanting.

For growers of rice, the time could scarcely have been more propitious to invest in the dykes and canals needed for transplanting. Households bulged with semiemployed. Within a year or two the unbroken land was honeycombed with paddy fields, only a few patches remaining where water flooded to depths of a yard or more, these given over to broadcast "floating" rice.

If, in accordance with Table 4.2, yields increased by a third in the change from broadcasting to transplanting, the weight of the grain at harvest rose in fields of 30 *rai* from 8 to 11 short tons, and cash in hand from about 150 to 200 *baht* ($75 to $100 at the existing rate of exchange). With slight upward variations from this bottom price, packed households sustained themselves through the thirties. Trips to the market were few. Cultivators fashioned their own irrigating equipment, hammered worn tools into shape, and set off for the fields in tattered shirts.

Then came the still grimmer years of World War II, when freighters ceased to cross the mud bar at the mouth of the Chao Phraya River, mills closed down, and grain moved no farther than the household storage bins. Everybody was grateful for some of this residue, when floods washed away the 1942 crop. Aside from occasional black-market deals, nothing was available to buy or sell, except what people from the city brought to the cultivator's doorstep, perhaps a piece of cloth to be bartered for some eggs and a sack of grain. A number of young men earned a few *baht* digging earth for the Japanese army and spent them in a few minutes for some black-market treasure, but most people, ashamed of their rags, stayed home. In 1945, with the cessation of hostilities, the year's crop appeared in slightly reduced abundance, but with bombed-out docks and crushed railway bridges, the grain could go nowhere. So the postwar boom almost sailed past most of Southeast Asia. In subsequent years, however, grain prices rose, for civil unrest in almost every quarter pared production. Thailand, almost alone in this section of the world, had abundance for export, more than ever before.

NEIGHBORHOOD AND COMMUNITY

The landscape has changed considerably over half a century. Like central Illinois in miniature, large dwellings dot the countryside, each on its own plot of land, with rice instead of corn or wheat, with canals instead of roads, dwellings roofed with tile or thatch instead of asbestos shingles, with the tails of nagas on roof tops instead of lightning rods. Now the rice fields have erased the occasional patch of bush where birds once nested and mosquitoes bred in the shade. Without birds children need no longer guard the ripening crop with their slingshots or with strings of flashing objects dancing in the wind. The smudge fires that protected man and buffalo from the swarms of mosquitoes have dissolved into cold ash. Canals are deeper and better defined, permitting the grain boats a longer season to take away the crop. Smaller ditches radiate out to bring irrigation water to the seedbeds. Little houses that were once starkly exposed to wind and rain have become larger, protected by luxuriant trees which also provide firewood. Once a single multipurpose shelter, the temple now embraces a variety of buildings, pavilions, pagodas, and ceremonial furnishings. An orange-roofed school with flagpole standing symmetrically before it is situated nearby. Especially striking are the newly-mounded highways where trucks and passenger cars add to the roar and exhaust fumes made by the outboard motors of the boats in the canal.

Pick out any spot along these canals. One no longer finds many areas where neighbors share common ancestors or even come from the same geographical area. Nuan, for example, is a thin, small woman in her mid-forties. She owns 30 *rai* (12 acres) of prime rice fields, and has never married. Born in Bang Chan, she inherited land from her parents. "Raising children makes too much work," she declares, echoing the plaints of many women who have, nevertheless, married and borne children. The two sons of her older sister have come from some distance to live with her and help work her fields; one of them is already married, having brought his wife to live in the ménage he may inherit. Across the canal is the land of Nai Thim, worked by a tenant who moved in two years ago. Up the canal an old man lives with his only son and daughter-in-law. Nuan cannot remember how many years these people have lived there, but they came after World War II. Chya and Thep, with 35 *rai* of their own land down the canal, have lived there as long as Nuan can remember. Thep calls Nuan "older sister," though in fact the two are unrelated.

Most Bang Chan householders now maintain a certain aloofness from their neighbors. Paddlers pause to converse in passing, or someone stops for a moment in the shade of a verandah, but the basis of reciprocity rests with an occasional cheroot or a drink of illicit, homemade liquor. At the store next to the temple men may discourse tranquilly on the level of water in the canal, the price of rice, perhaps about a neighbor, but rarely about their own frustrations and aspirations. A landlord like Nai Thim might be expected to develop trustful symbiotic relations with his tenant, who routinely borrows his buffalo and helps repair his house. Actually, the relationship has become mechanical, for people say Nai Thim is spending most of his time in Bangkok trying to borrow money from a rich relative. Labor exchange between neighbors all but ceased some years ago, for people found it difficult to correct a neighbor doing a job badly. Now everyone in that part of Bang Chan hires his extra labor. Nuan usually hires a man living in a thatch house beyond the school; having no rice land or agricultural tools, this laborer has rented a tiny space for a house site. Older inhabitants contrast today's Bang Chan with the period of hamlets when kinsmen lived adjacently: "We could eat in any house. There was always food enough. If anyone planted on my land, that made no difference. We gave money to anyone who needed it, as long as it lasted."

This new distance is also maintained by tacit assumptions of superiority without symbiotic liaison that a larger landowner feels toward a smaller one, or by the vocabulary of deference used when a tenant speaks to any landowner, not necessarily his landlord. The wage-laborer knows his place too, living in a makeshift house, eating less tasty food, and sending his children to school in patched uniforms. To be sure, each household, rich or poor, tends its own fields, sells its own crop to the agent from the rice mill on its own terms, gets rich, or subsides into debt by itself. Yet when Nai Thim sends out invitations to join in the celebration for his son who is entering the monastic order at the local temple, he shows one surviving vestige of dependence on his neighbors. Its form is adapted to the new life. Though no one is excluded, he invites particularly those deemed able to make a contribution to the expenses.

On the day of the celebration the name of each contributor, together with the amount contributed, is carefully written in a school notebook, later filed away for future reference in the drawer with important papers. These become the obligations of Nai Thim toward each contributor when that man, next year or a decade hence, stages a tonsure ceremony, a marriage, a cremation or some other life-cycle rite. The host alone knows the actual number of guests and how much was given altogether, but in their sampans returning homeward through the dusk, guests comment on the size of the crowd, the tastiness and quantity of the food, the kinds of entertainment, and the probable cost to the host. From there the conversational step is small to comparing Nai Thim's celebration with others of the vicinity.

Ranking rests only incidentally on the inventory of one's possessions, more squarely on the number of reciprocities in which one is engaged. The advent of money has shaped the mode of operating but not the objectives. Like any tenant or laborer, the rich man too is in debt, indeed more deeply in debt than they, for he uses his greater resources to borrow larger sums. Thereby he lends to local people, and helps his scattered kinsmen buy farms or buffaloes. His large house is filled with children and nephews whom he supplies with a *baht* for candy at the local store or 50 *baht* for a trip to Bangkok. The boats, portable gasoline engine, pumps, ropes, yokes, and six buffaloes sheltered under the house are also loaned out, helping bring people to him. He spends every available coin and differs from tenants only because he has spent more money, acquired more possessions, and hence is said to have acquired greater merit in past lives. He knows that his merit from the past has not sufficed to permit him rebirth as an orchardist, a Bangkok merchant, or a government official, but at least he was not born a laborer or a tenant. He manages his assets so skillfully that when creditors clamor for payment, he rarely has to sell a buffalo or boat, but can collect enough from several debtors to pay the demand.

How did these transformations take place? After the first five years of transplanting, most households settled down to using little more labor than they first required for broadcasting, but those five years were strenuous. Dykes had to be raised to hold the water that guided leveling the ground to a plane parallel with the surface of the water. Easygoing broadcasters had to learn the precision necessary for making splash wheels and dragon-bone pumps. Children used to turn the treadmill for hours to irrigate a seedbed, finally to be relieved by adults who continued working into the night. It took one day to flood a *rai* of land, a demand that became nearly intolerable when a rainless June required three or even four inundations to prepare and plant the seedbed, for 1 *rai* of seedlings was needed to fill 10 *rai* with transplanted shoots. Thus between 3 and 12 man-days must be devoted to the seedbed for 30 *rai* of land under cultivation. The advent of the windmill to power these irrigating devices offered welcome relief, for then a stiff southwest breeze could flood a single *rai* in an hour. Portable gasoline engines followed in the 1950s.

Cultivators learned to hurry with plowing, preparing all fields during the 35 to 40 days it took the seedlings to ripen for transplanting. Two buffalo were not enough for 30 *rai;* they tire easily. Tales drifted through Bang Chan about tractors, but the two or three who could afford to buy one did not even look at a machine, which, they had heard, might burrow itself hopelessly into the muck. Not until the 1960s did anyone try the Japanese substitute with skilike runner and caterpillar traction which can plow 5 to 10 *rai* per day.

Today they have learned to save their strength in plowing and harrowing. No longer is it necessary to plant all 30 *rai* at one time. Rice varieties with differing periods of maturation are sown, transplanted, and harvested in succession. Each variety comes due at its moment and requires few extra hands. Buffaloes still tread loose the rice, but now the 12-year-old girl who once poured a few pounds of freshly threshed grain on a rattan tray to toss it free from chaff takes her place at the wooden winnowing machine. She turns the crank that sets the fan and belted wheels whirling while her brother shovels the threshed grain into the maw. Chaff flies out one side, clean grain on the other, and the two winnow 10 times as much in a few hours as by the old method.

Three full-time workers can now operate 30 *rai* of rice fields by the transplanting technique, providing they have sufficient draft animals, the occasional help of children and aged, plus access to hired workers. Nevertheless, the problem of production remains command of workers. Janlekha (1955) describes the Chom family, with four adults and the occasional help of two adolescent daughters, cultivating 49 *rai* (19.6 acres); two years later when the two daughters grew strong enough, all worked 70 *rai*. Chom owned only 7 *rai* of land, but he rented more as his work force expanded. When his children left home, he rented less land. Tenants sometimes complain about difficulties in finding a landlord who will rent to them, but Chom, with his reputation for industry, had no such trouble.

Involved with the labor problem one way or another is the cash shortage. No longer does the cultivator approach his neighbors to gain services. He needs money for capital equipment, labor, and even to hold a household together. People like a more varied diet than the local gatherer can find at any given season. Today they wish tailored shirts and trousers, shoes and hats instead of loin cloths. For symbiosis in this decade cash must be nearly as available as a flagon of water or a sleeping mat. Otherwise people leave the household.

Cash production continues the year around. Gardens and patches of water on the canal's edge produce lotus, greens, pineapple, eggplant, and other edibles, while under the house poultry and eggs are raised. Almost every day some woman sets off with a basket or two of produce to sell in the markets of Bangkok. People buy a sewing machine to earn money by making clothing to order, or a portable gasoline motor that can be rented for cash up and down the canal. A young man hopes to make a little extra repairing radios, a skill learned in a Bangkok trade school. Through the dry season many seek jobs as laborers, and it helps if a grown child holds a permanent job, providing some cash returns to the parental household. Regular pay raises the desirability of marriage to a teacher at the local school or to one of the local maintenance men on the new highway.

Few vestiges of kinship clusters remain in Bang Chan. Up the canal is one where three households work land that the oldest brother received from his father. This man, an only child, married and then brought his wife's kinsmen to live with him. Where the highway bridge crosses Bang Chan Canal another cluster is found; there kinsmen still exchange labor at transplanting time. But postwar prosperity helped dissipate most kinship groups. It is hard to hold children at home, for eventually a daughter and her husband save enough to buy their own equipment and move an hour or two away by paddled boat to work their own land. Kinship clusters also dissipate when the parental generation dies, and their children decide to sell their equal shares in the land. When heirs require prompt payment, the land goes to the man with ready cash in hand, often not a kinsman at all.

Where kinship offers a basis for reciprocity beyond the household, it must operate at a distance. A daughter of a Bang Chan cultivator married a policeman who rose to become a colonel in the ranks. The colonel's wife still holds land in Bang Chan. This land is worked by her cousin and produces the rice needed for their Bangkok household. Once the wife requested the privilege of giving robes to the monks in a large ceremony at the Bang Chan temple, and since the occasion also implies generous donations to the temple, the abbot was very happy to receive these merit-makers. A well-known parallel case is the Bang Chan man who, helped by a friendly priest to enroll in a Bangkok school of note, rose to become a captain in the Royal Navy. His older brother similarly supplies rice for the captain's household from the "family" land in Bang Chan, and sends his children to live with their uncle while they attend high school or college. When the country brother built the new congregating hall at the Bang Chan temple, the Bangkok brother contributed substantially. Money and influence hold these siblings together in ways that disappeared from the local scene with the death of commune headman Phlym in the 1920s.

In most cases, kinship reciprocities have narrowed to include fewer people. Parents advance repayable but interest-free money to their own children but rarely to the children of their siblings. An older brother, perhaps employed as teacher in some distant town, invites a younger brother to live in his rooms and accept his guidance, but less often do we find cousins treated in this manner. Children in the army or working at some urban job return to help parents during a period of heavy field work, but only occasionally to help a sibling. Quite a few kinsmen can be drawn to a life-cycle ceremony, however; some spend two or three days preparing food, cleaning the house, and arranging floral decorations. More distant kinsmen ordinarily come for a day and make somewhat larger contributions in money than the ordinary neighbor. Failing to appear, or to send a little money, is a sign of ill-will, for these are times for solidarity.

Kinsmen living in Bang Chan, however, remain a potential step closer when one needs to seek help. A farmer finds it easier to apply for land from a landlord cousin, and may expect to pay a little less rent. Should such a tenant be asked some year to look for another place to grow his rice, the affront cuts deeper, coming from a kinsman. Similarly, one should be able to borrow money on more lenient terms, with less interest to pay and more flexible repayment, and a friendly sibling sends his 14-year-old daughter to help for a few days with transplanting, or loans a boat until the old one can be repaired—but of course a neighbor who has become a quasi "aunt" or "mother's younger brother" will do the same.

Though Bang Chan has become differentiated into individual households, most of them are not literally isolated. All have increased contact with persons outside the limits of Bang Chan. Brothers and sisters scatter all the way to Bangkok and beyond. As we shall see in the following sections, liaisons have developed with the commercial community as well as with the government and monastic communities. Bang Chan is no longer self-contained or geographically circumscribed. At its center we see only a segment of its life and must infer from the comings and goings by boat and bus that Bang Chan extends well into its surroundings.

THE COMMERCIAL COMMUNITY

Captains of junks out of Canton and leaders of caravans along the silk routes were notable representatives of the commercial community. These roving traders, with little more than their boats or their ponies, had to live on the uncertain difference between what they paid at home for silk or porcelain and what some distant customer might offer. Their precarious livelihood distinguished them little from the pirates and brigands who lurked along the routes, and when the original traders did not reach the next market with their goods, these less scrupulous types often took their places. Once at the market, any shrewd person could carry on; the task of making a deal was made easier if money were acceptable rather than barter alone being used for negotiation.

Other people, however, stayed at home, supporting themselves less hazardously by producing for their own consumption. Their rice nourished the household. If they wove cloth, they themselves wore it. Under most circumstances they had barely enough, certainly no surplus to trade. The multitude of people worked to subsist.

Chinese merchant groups bent on exchange, and Thai rice growers intent on sustaining themselves with consumable goods, lived side by side for centuries with little intercourse. Chaem and his brothers or Mau An and his kinsmen went a few times each year to the Bangkok market or a store on Saen Saeb Canal in order to barter fish or unhusked rice for a few needed items. Boys stole a bucketful from the rice bin and secretly spent it at a store for salty prunes. Little, if any, currency was seen, and wares stood long on the shelves. So it might have continued, had not World War I sped those buyers into the byways with currency in their pockets. Then all bought more goods more frequently, and made fewer things at home; yet they still lived with a subsistence orientation. It took the children of these people to understand that the market economy measures well-being in profit. Consequently they began to hold their crops, if they could, until the price should rise, perhaps to buy a portable gasoline engine, not solely for their own use, but to rent out to neighbors. Only a new generation could grasp this orientation, and only their children could learn the techniques of publicity, speculation, and market manipulation that are called "economic development." In the history of Asia, however, we note no dazzling metamorphosis but only the extension to the whole society of a viewpoint long associated with one community.

At the fringe of today's commercial community is the Chinese peddler who accepts a consignment of damaged goods from some cottage industry, or garlic and peppers from a merchant with no more storage space. The consigner risks little, and the peddler's risk arises from the terms of his sales, for in many a district he hands over the goods on promises to pay later. Before long, since much of his time is consumed trying to collect old debts, a man well decide that day-labor for wages brings better returns. Somewhat more secure is the petty merchant who can afford to hire a place in the market area, there to set up for the afternoon an awning to keep customers and merchandise cool or dry, depending on the season. If the market is well attended, he need only lure his customers with prices honed to a fine margin of profit. We may ascend the hierarchy of this community still further to the storekeeper, the factory owner, and even to the goldsmith-banker.

Here personal liaison to a financer is critical. Each entrepreneur has brought a few more assets with him and requires more generous terms of credit. The peddler receives little in advance. A petty merchant starts with more and arouses greater feelings of confidence, because of being a kinsman, because of some noteworthy service in the past, or because he can at that moment advance the interest of his financer. Between each entrepreneur and his financer can be found symbiotic liaisons that provide goods and services to fit the particular circumstance.

Unlike the monastic and government communities, no single hierarchy rises into view, even at a distance. Near the summits of these multiple hierarchies are clusters of kinsmen, often descendants of some father or grandfather from Canton or Fukien who worked his way up from humble beginnings. They, their sons and nephews, form a closed group seeking to invest funds and manage varied enterprises by placing some kinsman over each. Thus the rice mills in scattered provincial spots become variously linked to trucking firms, canal boat syndicates, warehouses, docks, and shipping lines in the effort to create a little empire.

According to the amount of capital available in any particular year, these mills buy rice. About November, former buyers of grain reappear at the office for another year, and each of these 20 persons more or less is proffered a contract to buy rice. The manager offers to buy maybe 500 tons at 800 *baht* from one, from another 300 tons at 775 *baht,* from a third 200 at 750 *baht,* according to the particular circumstance of the individual liaison to the manager. Some of them may bargain a bit, some receive advances of money. Then off they go from house to house among the cultivators offering to buy the crop or whatever part the grower may be willing to sell just after harvest. Their offered price is lower than that of the mill, the difference furnishing their profit. Some may make a little more by advancing money while the crop still stands green in the fields, for then the offering price can be cut a little more.

A few weeks later, after harvest, the buyer and two or three laborers appear in their long, narrow boats to take the grain and make final payments. On delivery at the mill, the contract is finished, and these buyers turn to other, probably less lucrative employment until the next crop is due. The small permanent staff carries on at the mill from there: a literate bookkeeper, a semiliterate custodian of the warehouse, a mechanic to maintain the machinery, a fireman, and perhaps a janitorial coolie, all of whom live in the mill compound. They may receive some salary, but their housing, their food, and their chances for extra compensation in buying some part of the crop or in borrowing the company truck for the afternoon usually constitutes compensation enough to hold them many years.

Well before the 1950s Thai rice growers and agricultural laborers sometimes tried their fortunes at storekeeping in hopes of earning a little money. In the absence of a financer they bought outright what merchandise they could with the money at their disposal and opened for business under a thatched roof beside the canal. Catering to local needs, they often had a pot of red lime for betel chewers, a cluster of metal mattocks suspended on a wire from a roof beam, a bundle of cheroots, candles and joss sticks for home ceremonies, and perhaps a few bottles of liquor. Sales were rare, though occasional neighbors appeared, sometimes with money, often promising to pay later. Then as accounts mounted beyond hope of collection, the hapless proprietor took to drinking up the remaining liquor with a few cronies. So another store succumbed, and rice growers felt their judgments confirmed that living from the market invited disaster.

Aun, however, was an exception. In the early 1950s he and his wife Bun sold their house and rice fields. Childless and too old for rigorous field work, they proposed to open a store at the bridge where the new highway crosses Bang Chan Canal. People moralized on Aun's folly up and down the canals, while friends quietly tried to dissuade him. Despite their advice he persisted. In a few weeks a monk came to the building site to bless the mother pole of the unfinished store. On the opening day for business, Aun and Bun made merit by inviting monks to eat and read a sermon. Then curious people stopped to see the bottles of liquor on the shelf, the packages of cigarettes and patent medicines, perhaps to taste the cooked noodles ready to eat, or to drink a cup of black coffee. Six months later, to everyone's astonishment, Aun was still in business.

No one of his advisers, probably not even Aun himself, realized that enough cash was then flowing through the hands of cultivators to sustain a store. They had heard only the complaints of money shortage without noticing the new sources. The young were no longer selling the whole crop at the nadir prices after harvest; many were holding some portion to take advantage of rising prices three and six months later. Women with baskets of eggs, mushrooms, and lotus stopped at the store returning from their trips to the market; they had money. Clusters of men on their way to and from work paused to sip a cup of coffee, and they received their pay each week. Everyone who climbed on the Bangkok and Minburi buses that passed Aun's door had a few *baht*. Indeed, many a bus stopped to let off young men and women returning from Bangkok with wages in their pockets and watches on their wrists. As soon as Aun's prosperity was evident, four new stores appeared near the bus stop on the highway.

The commercial community, now both Chinese and Thai, no longer lies at the periphery but has moved into Bang Chan. Across the highway from Aun's store is a small rice mill run by diesel motor, where housewives bring a few baskets of unhusked rice to be milled, instead of using the hand grinder at home. In the nearby stores stocks of edibles are to be found for daily consumption: bananas, fish soy, dried fish. The widow Chya has given her rice land to her son, so that she may run a small food stand near the temple. Householders like to vary their diet with a handful of salted shrimp or some greens from the market. Trucks with soft drinks and others with ice, flashlight batteries, or patent medicines come from distribution centers in Bangkok to replenish the wares. At the junction of the highway, half a dozen taxi boats with outboard motors await the incoming buses, and for a few *baht* will roar a passenger up the canal to his destination. En route one passes households where young women are sewing blouses and shirts by machine. A little farther along some young man has set up a barber chair and also repairs radios. An agricultural laborer in the next house has set the members of his household to making a type of basket which, attached to a long pole, helps pick orchard crops. Of course, anyone wanting to buy a pair of shoes or a yardage of cloth must travel to Minburi or Bangkok. More than 80 per cent of the households still raise rice or work as laborers, and probably this percentage stands close to the income of the area represented by agriculture as compared with the income from wages, shopkeeping, etc. However, this once uniform rice-growing community is interlaced with commerciants whose patron-financers reside in the remote periphery.

ORDER AND MERIT

The government community has also extended its presence in Bang Chan. Once represented solely by the hamlet and commune headmen, new agencies have come into the area at several points. Three police stations are located within a few miles' radius from the Bang Chan temple, and one may see a boatload of these men in gray uniforms patrolling the canals every few weeks. They make few arrests, partly because few local people have money enough to activate the policeman's license to arrest people and "protect the public." Hence most people manage their own problems. When a man has lost his boat, he lets his plight be known as broadly as possible. In two or three days someone approaches him to negotiate a price for redeeming the missing property. After a few days a settlement is reached, the property is returned, and the owner resolves to guard his possessions more carefully. No police are involved.

Similarly, the corps of agricultural officers, veterinarians, sanitarians, midwives, dentists, and physicians attached to the district office have few contacts with Bang Chan. Rarely do these petty officials come visiting, and local people, seeking them out in their offices, feel the necessity of an introduction by some mutually known person in order to avoid the usual arrogant indifference. Most people prefer to discuss their problems of rice growing with a successful neighbor. They take their sick to local practitioners of traditional medicine; should treatment fail, only then do they turn in despair to government clinics and hospitals.

At the district office, one pays the head tax, secures the license for peddling or storekeeping, registers a birth, death, or marriage, and each time hands over a few extra *baht* for the clerk to pocket. Young men report to these offices at selection times for the military service. At the nearby district court the more implacable neighbors seek to gain justice for their disputed claims over land and money, though ordinary mortals, less tenacious, will have worked out a compromise many months earlier aided by some local go-between. People rarely seek out government services.

Government has, however, moved into education, displacing the monks who for many years taught boys to read sacred scripture. The new teachers are hired by government, wear their government uniforms on official occasions, but differ markedly from most officialdom in living among rice growers rather than at the district office. They make merit with local people, send their children to local schools, chat over "Pepsi" at the local store, and often grew up in the very area where they teach. Though their pay is modest, and their opportunities to acquire extra income limited, they are the most affluent persons in the area, because of their regular monthly pay checks. Local people consider having a son or daughter employed in the school system, or married to a school teacher, a definite advantage.

The value of the service that the teachers offer is another question. Some growers ponder the desirability of teaching children to read books that have no moral content. Others grumble that teaching girls to read and write only aids them to pass love letters to young men, that a farmer really does not have to read or write, and that classrooms use time that might better be spent tending the buffaloes. Yet others observe that with some knowledge of arithmetic, one can determine the yield of a rice field and determine the cash return from selling seven tons of grain. Some pupils have been able to help support their aging parents by learning enough to enter a government job. The pleasant is somewhat mixed with the unpleasant.

Whatever coolness government generates in the minds of most people is readily dissipated as soon as a local project is announced. For instance, when silt has reduced boat travel to a few months of the year, Bang Chan becomes enthusiastic about a proposal to dig out the local canals. Then grain and other produce can go to the market at any season; one need not walk with heavy tools on his shoulder; water is ever there for transport, bathing, and cooking. When the provincial governor receives a sum for such a project, he too is happy to estimate the needs of all the districts within the province and pocket a portion for his own use. At the district and commune level, each official in turn accepts the fraction passed on to him, then passes on what is deemed sufficient to carry out the work. In the end, the hamlet headman divides the work assigned to him among the residents, each being paid for the number of square yards of earth that have been moved at a rate fixed between the headman and the resident householder in question. To be sure the work might have been accomplished at two-thirds of the price, had the governor of the province hired his own workers to do the entire job, but he would have had to sacrifice the benefits of a reinforced hierarchy within the government community. Instead of paying constant wages to hold an organization intact, an inadequate wage is supplemented occasionally by reassuring bonuses. As long as no one takes too large a share when the sum flashes past, schools and public buildings arise, and highways are built. When an official pockets so much that the job cannot be completed, then one may speak of corruption rather than of an appurtenance of office.

As the government community enters the local scene more intrusively, the monastic community, though still fixed in the old temple compound, recedes a step. Few children now serve the monks as temple boys, but paddle past the temple without stopping on their way to school. People assert that just as many young men as ever enter the monastic order. While this may be true, their months of devotion must be squared with such competing obligations as the rice crop, work for wages, training courses, and military service. What has disappeared are the dozens of adults who once spent sociable days after harvest at work in the temple, repairing the roofs, assembling a house to quarter the monks. They are now busy earning money, and the temple too must tend to its needs by earning money.

The annual fair at the temple with its boxing match, theatrical troupe, and the sale of gold leaf for those who would enhance the glory of the Buddha's footprint and their merit with a patch of gold, these and other ways of earning money net a thousand *baht* by the time the bills have been paid. In addition, the annual ceremonial presentation of new yellow robes to poor monks by workers in some government department has become a fashionable occasion for excursions to the countryside, and brings generous donations to the temple. Even for rice growers the offering of flowers once picked from the canal banks and presented to a monk in a bowl with food, candles, and incense sticks has been replaced by the bristling money tree, its branches made of quivering wire, its leaves of one *baht* notes. The very symbols of devotion have become money and what money will buy. And yet some merit-makers still occasionally hire a worker to mend the roof and shore up a rotting post, lest the temple fall into ruin.

The users of the temple have become more broadly differentiated. The little group of kinsmen still appears at the sanctuary door with a young man on the shoulders of his older brother who presents him for acceptance into the monastic order. They pass nearly unnoticed beside the noisy band with musicians and dancers that have come with the son of some prosperous household from a day or more of feasting. A humble man can still quietly cremate the corpse of a parent at the pit behind the temple. However, anyone who aspires to standing cannot avoid rental of cremation furnishings from Bangkok, the hiring of cooks to feed the invited monks and guests, the negotiations for a gasoline-powered generator to illuminate a dramatic performance through the night, and the purchase of rockets to announce to the heavens, when fires have been set in the cremation pavilion, that a soul is on its way.

The wizened abbot, worrying lest his temple be by-passed by the wealthy merit-makers for more resplendent neighboring temples, has turned actively to refurbishing. During the Depression and war decades, the floors in the monk's quarters were hastily patched where boards had given way. The sanctuary, barely 40 years old, had settled, leaving large cracks in the brick walls and stucco ornaments crashing down from the cornices. The old congregating hall, never more than a jerry-built structure of timbers roofed with corrugated iron, became ever more incongruous in a period of status consciousness.

In the late 1940s, the abbot drew together a committee of prominent local people to survey the needs, and soon a trim, new congregating hall arose with glittering nagas on its roof. The naval captain in Bangkok and his local brother made this merit. Soon other elders contributed a building to house a few monks, a new bell tower, and a cremation platform, but the sanctuary constituted an increasing hazard to merit-makers. Enquiries about cost of replacement produced astronomic figures which daunted the lay members of his committee but only renewed the abbot's determination. Clearly no local cluster of kinsmen, even when backed by wealthy uncles and aunts in Bangkok, could be found for such an undertaking. If, however, a sum could be collected little by little over the years, if it could be swelled by loaning at the usual interest rates of 10 to 25 per cent per year, and if his merit sufficed to live that long, the abbot might still see a new sanctuary.

Fortune smiled on the project, not only because of local contributions. Bus travel and creeping urban growth reduced the isolation of once distant rice fields. The abbot capitalized on his visits to superiors in the monastic hierarchy, brought them to visit his temple where they could see the peril with their own eyes, and drew on their merit-making inclinations. For more than a decade he worked, and by the late sixties hired workers were pouring concrete for the foundation of a new and greater sanctuary. At this moment the abbot rivaled even the district officer in work projects for the local people.

The temple, long accustomed to live from the hearts of its devotees, now commands people as a patron of local works. The homemade pavilion installed with a local householder's devotion has become a red gilded construction ordered from a Bangkok factory, delivered to the temple, complete with a brass plate bearing the donor's name.

ADAPTATIONS TO A SOCIETAL ENVIRONMENT

These were Bang Chan's years of adolescence, but adolescence is not just nubility. It also implies extended perspectives on the world and the self. For Bang Chan the transforming element was cash, the result an infatuation with the market. But out of the narcissism of consumer buying there arose producers of rice for unseen persons in distant lands. Freed from a single preoccupation with the rice fields, Bang Chan's skills expanded in dealing with machinery, electricity, and the written word. People traveled to scarcely-heard-of places, earned pay using never-before-recognized skills, and became aware of a world with lands like Korea, Africa, and Russia. Like youth asserting its independence of parents, Bang Chan by-passed its nearest neighbors in search of companionship with distant members of the commercial community. Awesome persons in the government community no longer awakened old fears, while old symbols of security from the temple, once pure and simple in their meaning, became confused with social status.

Bang Chan of the transplanting period has been sketched in the metaphor of adolescence. Our task, however, is to describe these changes through another metaphor, adaptation to environment. We may not portray change as coming from within Bang Chan, nor imply that the changes form an oft-repeated sequence leading step by step toward maturity. Instead we shall see changes as a manifestation of altered relationships between households and their environment. These adaptations to new environmental conditions may be local or historical only; they follow no special direction but sustain in some manner the functioning integrity of these households.

More than we anticipated in our comparisons with broadcasting, the change to transplanting demanded a heavy increase in labor. The heavier labor investment occurred first in the capital requirements for preparing irrigable fields and, second, in the management of the crop. The latter demands were first mitigated by drawing on the power of wind to irrigate, and on more buffaloes to plow. Gradually people added boats for hauling and machines for winnowing more conveniently the larger crops. They further learned to plant several varieties of rice in sequence with differing rates of maturation, thus spreading the peak demands to manageable proportions for a household. Some sought to increase yields additionally by applying chemical fertilizer, others by trying new rice varieties. In particular, however, these people learned to expand and contract the areas under cultivation; even when property lines might be expected to limit flexibility, renting to neighbors gave the scope needed to sustain a constant energy input over the land as a whole. Through such means Bang Chan was able to sustain the increased energy demands of transplanted rice.

If we may assign causes for this change, primary was the new relationship to the societal environment. The eager rice buyers of post-World War I sought out the growers, who accepted the proffered money and stood ready with more rice for sale on the following year. More troublesome than raising more rice, however, was the management of cash. At first cultivators became thirsty consumers of goods, not just because of the lure of the marketplace but because holding onto a hundred *baht* as a boat or a buffalo seemed safer than keeping fragile paper that might at any moment be burned, torn, lost, or stolen. As markets moved closer to Bang Chan, the value of access to money was evident, and property became less a utility for the growing of rice than a symbol of wealth convertible to cash. Yet soon they realized that sheer accumulation of livestock and tools guaranteed no safety in a period of declining prices. Then some took the next step, buying goods that produce income, the gasoline motor that can be rented out, the sewing machine that produces dresses. At that moment they understood a key concept for survival in a market economy—not to buy for consumption, but for the margin of return over cost.

Household heads had to learn new ways to retain their labor. The yardage of cloth that once satisfied a household for a year required generous supplements with canvas shoes and store-made shirts. While "modernizing" in these and other respects helped provide an atmosphere of well-being, to make cash available proved more complex. Household heads needed a few *baht* on hand for the child at school, a trip to the store, or the young man who wished to spend the day in Bangkok. At first all shared the coins from a jar, as if they were water or rice. When demands emptied the container, household members sought to establish the principle that sums removed must be returned, a device that did not long survive. Then they encountered days of plenty and months of cashlessness. Soon they discovered that the housewife with garden produce to sell was a valuable asset, and when even the busy trader at the market had no money to advance, many people turned to spending a few weeks at wage work. Yet this also became a danger, for requiring a young man to turn over his earnings to the household in the old way became an ineffective means to hold labor. By the 1950s fathers were working out partnerships with the children they hoped to retain. A son-in-law might be lured to stay by jointly working the entire land, but, at harvest, keeping as his own the crop from certain fields. Such a device rewarded young men with cash beyond what they could get from wage work, thus holding them to the household, without invading capital assets in land or tools. Moreover, the share, ever subject to increase or decrease, allowed flexibility for possible future bargaining.

With benefits centered on retaining a household membership that might turn fickle at any moment, household heads relaxed reciprocation with all but selected neighbors. The hospitality of the kitchen could not be shared with its former ease. Though exchange of workers with neighbors saved money at transplanting and harvesting time, hiring laborers insured better quality of work, could be timed more readily to the state of the crop, and left one's own days free to seek wage work. Only when one needed the contributions of many to stage a life-cycle rite were old modes of support revitalized. Then a few *baht* put out to help a neighbor send his son to the monastic order or stage a wedding feast for a daughter would be returned on the day one became host for a life-cycle rite at one's own house. With 50 to 100 neighbors contributing, these exchanges offered reassuring support.

In relation to societal environment beyond the area, Bang Chan had to learn new types of reciprocity. When mother's younger brother demanded payment on an interest-free loan and when one could only find the money to repay him by borrowing at 50 per cent interest from the goldsmith in the market, a day of reckoning could not be long postponed. As they learned the advantages of credit and the dangers of foreclosure, owners became tenants, and tenants became laborers. The most bounteous harvest was of no avail against this kind of disaster. Cornered growers responded, as rice growers must, by growing more rice. In the new market economy this was the moment of conversion to transplanting.

In dealing with the commercial community, one began with a simple exchange of goods for a given price, as when the owner of a stall in the market bargained with anonymous customers he would never see again. The Bang Chan woman with her baskets of vegetables and fruits might appear in the chill dawn to sell to this keeper of a stall. Then, however, they sat and chatted about traffic on the highway or asked for advances and postponements of payment to serve some personal convenience. Even Bang Chan storekeepers achieved this degree of symbiosis with many of their customers whose debt was marked on a slate, presented annually, and finally settled through negotiation for an amount smaller than the written total. It was always more important to hold the customer than to lose a few *baht*.

Many rice growers merely waited for the buyer of rice from the mill who offered the highest price, but others waited for the wrinkled old Chinese with a peaked sun hat who had bought grain for decades from this same household. He offered his old customer a favorable price and readily advanced interest-free payment for future grain. Then a few weeks later he returned with a helper to load the grain. Before departing, the old Chinese scooped a handful and returned it to the grower saying, "Here is the soul of rice. Put it back in your bin. May next year's crop be larger than ever before!" Between the commercial community and the rice grower this exchange provided a strand in an interaction that ripened as apprehensions dissolved. Like other world religions, Buddhism preaches that more is to be gained by love than covetousness. So relations to the commercial community grew through the years.

Relations to the government community took another course. Bang Chan saw chiefly collectors of taxes, extorters of petty fees, and conscriptors for military service. Various rice growers voted in two or three elections for a representative at the national Parliament, yet when the candidate marched people off to vote for him, his requests rang like those of the police captain and the deputy district officer: favors done with little to show in return. Though all recognized the periodic work projects that benefited the participants, and the relief given when flood or drought decimated the crop, the net response remained ambivalent.

Where possible, Bang Chan evaded the government community. Land areas owned by a given person were generally underreported to save money. Changes of title often went unrecorded for years to save a registration fee. The birth of a child was registered a few years late, both to save the fee and to postpone the day for induction into military service, perhaps to hold the young man a little longer at his household job. Licenses to peddle goods or to run a store were obtained only when confrontation with the police seemed unavoidable.

Hamlet headmen attended the monthly meetings at the district office, listened politely to the district officer's latest instructions, and knew their duty to relay the messages on rural development to the people of the hamlet. Yet local meetings on these matters rarely occurred unless an official tour of inspection were imminent. Then headmen and a few of their closest dependents quickly built a sanitary privy or dug a fish pond for *tilapia* in order to satisfy official expectations. On the day of the visit, headmen rounded up the best food of the area, fed the guests well and sought to explain the poor showing: "People were too busy this year and too poor."

Bang Chan accepted the government community very much as it accepted the occasional cobras that crawl in shaded paths and lurk along the canal edge. If one were normally wary, nothing much happened in meeting them. We have seen a man seize a live cobra with his bare hands and hurl it dextrously over an embankment. Though once a year someone was bitten, no campaign has ever developed to rid Bang Chan of snakes. They are useful, as the low rodent population testifies. Moreover, a great cobra spread out its hood to protect the Lord Buddha from the elements, while he was meditating in the forest. Just as paddlers spot the ripples of a snake ahead of a boat in the canal, so Bang Chan has learned the watchfulness needed to live with government.

The monastic community has lived long by depending alternatively on the initiative of the people and on active propagation of the faith. According to Buddhist scripture, the first chapter of monks lived in the forest, fed by the unsolicited offerings of merit-makers, yet the Lord Buddha himself as well as his successors preached to kings who responded by building temples and monasteries. In Bang Chan, too, some abbots quietly waited for merit-makers while others actively solicited contributions. It took little urging to draw the old to the temple—they were concerned with their position in the next existence, and the scores who came on feast days seemed bound by custom. Nearest of kin assumed that they must feed their sons who had taken up residence in the temple, but to repair and build a more magnificent temple in the days of a cash economy required special efforts. Not to allow the temple to be by-passed or forgotten and slowly to crumble demanded effort, organization, and funds.

Some of these steps caused no difficulties in Bang Chan. Abbots have long developed circumlocutions for requests of help by simply indicating a broken fan, pointing to a collapsed roof, or lamenting with a laugh the absence of able workers. To gather lay helpers was easy. However, in entering the monastic order a monk pledges neither to touch silver or gold nor to buy or sell, but the rule has received elastic interpretation for years. In Bangkok, monks may be found buying goods in the market, even when unattended by a yellow-robed boy who hands out coins and accepts goods at the direction of his mentor. Thus the abbot of Bang Chan found ample precedent for dealing with money, particularly if it were paper bank notes rather than gold or silver. All agreed that for managing funds the abbot was an ideal person, little inclined to abscond.

A thornier question arose when the abbot sought to make the fund grow by loaning at interest, for, unlike Christian churches, Buddhist temples may not hold property to support themselves.

Fortunately money need not be considered property, and as the abbot planned to spend it improving the temple, he did not actually "hold" it. Moreover, people came asking for money and promising to return larger sums at a given date; this could be considered merit by the borrower rather than usury by the loaner. Thus the monastic order adapted with little pain to the market economy.

After food and other necessities were cared for in the household of a rice grower, merit-making followed, its extent reflecting local prosperity, communal as well as individual. In the local isolation, life-cycle celebrations and gifts to the temple became a means of demonstrating attainment, of asserting in a slightly competitive way a household's standing in local circles. In Buddhist thinking, the prosperity of the household was reinforced by merit, and the newly-made merit in turn reinforced prosperity, like giving strength to the strong. Local merit-making alone, nevertheless, did not suffice to refurbish the temple. Only by extending the circle of merit-makers to people outside Bang Chan was the orange tiled roof of the new sanctuary built. In turn the abbot became leader for a local work project which brought more cash to the community and more merit to the temple.

In this period Bang Chan was seeking to establish a new holding in its societal environment. The older, semiisolated years of Chaem and even Mau An required little energy of input with its intermittent bartering for goods at the nearest shops along the Saen Saeb Canal. Hamlet and commune headmen alone represented the government community, while the local monastic community lived passively among the rice growers. The new societal environment after World War I demanded greater input, stretching the resources of the larger households. More man-days were required to keep abreast with the market, more for a complicating local government with a variety of new branches, and more for the temple as well, if energy still remained.

At this point let us again return to "holdings" (see Chapter 4). Through holdings man constructs and maintains a system that adapts him to his social as well as his natural environment. Indeed, a changing social environment rather than natural forces provoked the rice growers to move from a subsistence to a market economy. They then grew rice more abundantly, later increasing the complexity of their holdings by adding a kitchen garden and more careful poultry raising in order to participate more fully in the societal environment. Besides they added to their holdings such skills as the management of money, such equipment as shoes and wrist watches, such capital investments as learning to read. Without these new dimensions of their holdings they cannot survive in today's Bang Chan.

Clearly the dimensions of these holdings that relate rice growers to society have increased in complexity and hence in demand on labor. As transplanting requires more work days than broadcasting, so a market economy demands more of its participants than a subsistence economy. Similarly, a householder who would stall off tax collectors and support a local temple must devote more hours or days to these tasks than grandfathers who never paid taxes or merely invoked guardian spirits at times of crisis.

In considering agriculture, we observed that the labor input must match the complexity of the holding, lest the crop fall short of requirements. This generalization appears equally pertinent to social transactions. The man who passes his neighbor with nothing but summary greetings is unlikely to be able to borrow his buffalo. The lover who invests little care in courtship, and ignores his overweaning rival, is unlikely to marry the bride, while the recluse unknown to local officials gains as few favors as a sycophant. Each transaction with society presumes an appropriate input that can be roughly approximated in labor.

During the last half-century Bang Chan seems not to have achieved holdings of appropriate complexity. Inputs—whether in fertilizer to grow more rice, in schooling to cope better with urban people, in peddling wares to earn more money, or in meetings with local officials over canals, taxes, or health—seem rarely to have resulted in adequate returns. Each transaction has led to renewed demands for more hours of labor. In contrast, the transition from broadcasting to transplanting was satisfactorily achieved within a few years, when crops became relatively dependable. Can Bang Chan find a holding of adequate complexity to survive in this more demanding social scene?

The strain on householders to retain labor is evident. Children yearn for wage-paying jobs in the city in order to buy the latest wares from Hong Kong and Japan. In slack times they are off to

Bangkok instead of fishing or helping repair the house, so that their time and earnings profit the household less than formerly. Women who once made their own sugar or wove mats have abandoned these tasks in order to sell a little produce in the market. Then they can vary the restricted diet of seasonally and locally available edibles with meats and vegetables bought at the market. In the meanwhile household heads seek extra funds beyond the crop to pay hired labor or embellish the establishment with labor-sustaining portable motors, pressure lamps, and other consumer goods.

Can holdings of adequate complexity be found to bring a suitable return without further draining the labor capacity of households? Thailand has been witnessing its own metamorphosis, the result of readier communication with distant countries that has brought a new economic system, technological changes, and, concomitantly, demands for more people to operate and regulate these novelties. The comparable transition vis à vis a natural environment might well be the first Chinese to transplant rice. We can imagine the decades of their trials and errors in constructing ditches that held water, discovering soils that retained it, inventing devices to lift water, determining the proper spacing of plants after transplanting, and ascertaining the proper depth of the water for each stage of growth. Such a period seems comparable to the one faced by Bang Chan in its current search for holdings of adequate complexity in relation to its social surroundings. The gluts and dearth of the market, the wars and reorganizations of government are no more predictable than the droughts, floods, and crop diseases of the agricultural world. The ravages of any army plowing tanks through a rice field are little more devastating than a wave of rats.

According to Buddhist doctrine, man's success in meeting menace depends upon the extent of his merit. Those with sufficient merit will triumph while those who are lacking will be overcome. This merit, however, gives no automatic release from suffering, for as a human being in this world, every man is a Job. Following the blessed way during times of stress tests one's will to refrain from stealing, lying, and killing, perhaps to avoid fornication and intoxication as well. Thus one is well advised by the Thai poet Khun S'ra Prasot in a portion of his "Song for

Sowing Rice," loosely translated by the present writer from the German of Draws-Tychsen:

> May no sprout fall
> to right or left.
> May we as well
> Whoever we are
> Follow that way.
> May no day come
> Without a sun,
> No Thailand be
> Without its rice.
> Give us the strength
> To follow that way
> Eternally.

Chapter 8

BANG CHAN IN THE 1970s

No one knows where adaptation leads. No one could have predicted that fish occasionally stranded on land would develop lungs and evolve eventually to reptiles. The man looking at the ripening crop waving in the wind about 1960 or 1965 would not have predicted the demise of rice growing in Bang Chan. Signs to the contrary were abundant. Multiple cropping had already begun, not in rice but in the small market produce that was inserted between rice harvests. People thirsted for money. To be sure, certain difficulties blocked the way to multiple cropping of rice, the most serious being the scarcity of water between January and June. Eventually, however, it would be feasible to pump ground water into dyked fields, as had already been done on an experimental basis in another part of Thailand's central plain. A less expensive device was the small, earth reservoir of a type known in Burma centuries ago and recently constructed in northeastern Thailand with the aid of bulldozers; the water accumulated during the rainy season irrigates fresh crops when the rains have stopped. Most people looking across the fields would have predicted that extensive multiple cropping would come when growers could operate in a more complex holding.

Witnesses of the scene would discover next that rice production in Bang Chan declined as much as one-third in some years during the 1960s. The immediate cause was "Yellow Orange Disease," known in the Philippines as "tungro." It appears soon after transplanting and discolors the slender leaves of the rice plant, with the result that the full weight of pendant grains never develops on the inflorescence. Like many diseases, this one strikes where a species density is heavy, so that the entire central plain was more or less afflicted. Its active agent is a virus borne by green leafhoppers. Though no cure has been discovered, certain rice varieties do exist that resist the disease. The efforts of the Ministry of Agriculture to grow and disseminate these new resistant varieties were, however, postponed by governmental priorities established by military rather than agricultural considerations. With no special funds available to counter the disease, it spread widely.

Initially most growers believed they were dealing with another passing affliction, but as it reappeared year after year, people tried variously to stem it. A few applied fertilizers and were blamed by others for so weakening the plants that susceptibility to the disease increased. Insect repellents failed to keep the hoppers away, though, of course, chemicals were never sufficiently available to spray an entire field. A grower hoped his varieties were resistant to disease after a symptomless year, but too often they proved infected the following season. Because of a substantially reduced rice crop, people sustained themselves with wage work and by selling garden produce.

The disease persisted but did not devastate, neither dispelling hopes nor forcing decisive action. Discouraged people plodded along, yet by the end of the decade many of the young people held permanent jobs for wages. The older generation continued to plant, their acreages reduced in size because of insufficient manpower, and often sold all but a pair of their buffaloes. Some returned to broadcasting.

The decision to abandon rice growing was relatively painless in Bang Chan, where the commercial and government communities had already established their presence. We may recognize this presence most easily as the sprawl of an extending city. Though the old royal palace remained as ever on the Chao Phraya River, the urban periphery expanded each year. Stores of frame construction heralded its advent: through years of repairing Aun and Bun's store and its neighbors of the local market, the thatched roofs and mat walls were replaced by the tiled roofs and wooden siding of urban shops.

We may think back to the 1890s when the outskirts of Bangkok were extending to Khlaung Tej, today's dock area, and when Mau An with his kinsmen were displaced from their rice fields. Eighty years later the grand- and great-grandchildren of Mau An were being displaced from Khlaung Kred. Land speculators reappeared, like vultures at the death of an animal. Discouraged rice growers were easy to dispossess, and profits reasonably certain, with commercial activity in the vicinity. Not 6 miles away a whole village was built for athletes at the Asian Games of 1967, and was subsequently sold at steep markups to Bangkok residents pinched for housing. Nearby an industrial park was attracting Japanese and German firms to build factories. Along the highway near Minburi a similar project was contemplated, so that in Bang Chan the signs of the urban periphery were unmistakable. Besides the wooden stores and the diesel-powered rice mill, bottling and canning plants appeared along the highway. Here and there former rice fields became the site of a suburban home surrounded by flower beds, lawns, and barbed wire. Elsewhere roads led off to clusters of smaller residences, while the remaining rice fields lay a few hundred yards behind this urbanizing strip centered on the highway.

There where the breeze blew urban sounds and smells, agriculture wilted. Manning factories and serving residents of new stucco houses easily drew the young from the fields. The new holding demanded greater inputs than agriculture and returned greater yields of currency, for anyone can see that the working season of most factories outlasts the growing season of rice. The team of workers supplies an annual input of a size greater than a household can muster. Households then became the feeding and sheltering adjuncts to factories, with old rice fields the dwelling sites for workers drawn to the scene.

Had we observed from the streets of Bangkok rather than the rice fields of Bang Chan, this engulfment by a city would have seemed quite predictable. An urban rather than an agricultural sequence appeared, yet we might still be unable to say exactly how far the tide would reach. Had the Yellow Orange Disease never come, a richer Bang Chan might have resisted conversion. We might have encountered in the 1970s local truck gardens with vegetables grown on earth heaped up above the flood level of the rainy season. Even if sale of land in Bang Chan were inevitable, solid households still intact might have journeyed with their possessions like most of their forefathers to new rice areas. Becoming a factory worker or a domestic servant was not inevitable.

SUCCESSION IN BANG CHAN

In natural history the word "succession" ordinarily means the sequence of species that tends to follow one after the other, such as the kinds of trees that occur in a forested area year after year when the glacier is retreating. The grasses at the foot of the glacier give way to pines and spruces as the temperature rises. The latter in turn tend to be replaced by beech and birch, followed by oak and hickory. What kind of sequence can be described for Bang Chan? We shall first examine certain regularities in the sequence of holdings; second, the agents of change; and third, some of the contingencies upon which these changes rest.

Regularities. The three modes of cultivating rice described in Chapter 3 did occur in Bang Chan, yet they served more as points of orientation than as the actual steps in Bang Chan's succession of holdings. Shifting cultivation ended not with the advent of buffalo and plow but in some obscure manner. When the new broadcasters entered, the older population of shifting cultivators either remained in the area, eventually acquiring the tools and livestock needed to cultivate as broadcasters, or drifted as shifting cultivators toward the free land at the periphery of settlement where their poverty better matched their neighbors. Similarly, the transition to transplanting from broadcasting occurred in two steps, an expansion of the area under cultivation and then the more expected shift to transplanting into dyked fields. Behind the modes of cultivation lie small changes of holding, and each represents an increment of energy input, so that the holdings represent a cumulation of complexity.

The hero on the stage throughout was the versatile household, though villages and hamlets entered and exited like secondary characters. The household put in its labor and consumed the output from the rice fields. Independent households of these people held their own against government, befriended the commercial community, and made merit at the temple. The very looseness of their organization gave strength. Unlike the rigid occidental family, such a household can expand and contract its personnel. In abundance it expands, in hard times contracts, assimilates kinsmen or strangers, links with others or remains alone, and rises anew after the death or departure of its head. By dint of its versatility it appears everywhere, performing any task from agriculture to manufacture within the limits of its strength. Only in the final scene, when work loads set by machines exceed its capacity, does it succumb as a working unit to factory-type organization.

Though energy inputs increased through the succession of holdings we found Chaem no busier than Mau An, and Mau An no busier than Aun and Bun in their store. All worked with little leisure from morning to night, filling their hours, days, and years, but the size of households increased. Each new generation was usually better equipped than its predecessor and had access to greater reserves of labor during the critical periods of the growing season. Moreover, efforts became more focused. While Chaem hunted and fished as well as planted, Mau An's group spent more time growing food. By Aun and Bun's day, people were almost exclusively agriculturalists, growing rice in one season, garden produce in the next. Energy increased then not only through tools but through specialization of the formerly less differentiated households.

The kinds of work also changed. Though rice was planted each year, new tools made it possible to carry out old tasks more quickly, but each new device—the plow that speeded field preparation or the dyked field that made control of water level possible—required more time for maintenance. The buffaloes had to be housed and fed; the dykes reinforced or rebuilt. With these new tasks sensitivities also changed, so that the eye that could no longer distinguish the track of a deer from that of a wild boar could detect in an instant that the photograph on the wall hung a hair away from the vertical. The ear that could instantly recognize the sound of a weaver bird now has become atuned to the coughs of a portable engine or the verbal intonations that raise or depress social station by a fraction. Such sensitivities distinguish the country cousin from the city cousin, the boor from the sophisticate, each in his particular niche.

AGENTS. Let us examine the agents of change in this sequence of ever more demanding and confining holdings: first, technology, which has sometimes been dubbed the result, sometimes the cause, of change. Doubtless the setting of Bang Chan with its annual floods prompted the use of boats for transportation, the desirability of nets and traps for fishing, and the planting of crops resistant to flooding. In turn each tool provoked certain changes, the increasing attention not only to maintenance, but to the search for additional tools, such as the windmill, and the dragon-bone pump that is only useful with dyked fields and irrigation canals. However, we may also note that shortages of energy were met not only by tools but by reducing the land under cultivation, by marrying a child off to a spouse that would live in, and by offering shares in the crop to a child who threatened to move out. We emphasize a continuity and interaction with technology and societal arrangements, for energy-amassing inventions are social and economic as well as engineering matters.

In addition to technology, land shortages and overpopulation have been deemed significant agents of change. We observed overpopulation in the aftermath of World War I, when tenants were dispossessed by their landlords. Yet this classical shortage, pictured by so many social theorists, is only one of many. The Laotian freedmen clung together as long as a shortage of buffalo and plow remained, and only when new plows and buffaloes became available did the village break apart. Thus shortages bind people together as well as drive them apart. To the pressure of overpopulation we add the pressures of underpopulation, for the major limitation of an area under cultivation has been the availability of labor. Labor exchange between neighbors in Bang Chan was a symptom of underpopulation, for it disappeared as soon as hired labor became available.

The foregoing example of growers dispossessing their tenants further exemplies the classical rift in interest known to Marxian analysts as the class struggle. A generation later the children of these owner-growers were dispossessed by the new industrial elite that sought to profit from land development. The antithesis of this factor may be called class amity, and it too helped form new holdings. Not only did Mau An and his kinsmen welcome the tenants to help clear land of brush, but these tenants welcomed the offer of land and rice for the season. Chaem's nephew, the commune headman, sold and rented lands to grateful kinsmen who brought more land under cultivation. The Chinese mill owners too aided their impoverished kinsmen by providing housing, and while the work they did was of advantage to their wealthy cousins, their advantage was not exclusive.

Those who wish to document the responsiveness of rice growers to change of price in the market will remember the jump in productivity after World War I. However, they may not forget that productivity also increased when prices dipped in the thirties. The increase in output at this time came from joining a system that increased specialization at the same time that it released households from the many hours of labor needed to make fish soy, sugar, mats, and hats. Money intervened, facilitating exchanges that converted rice into many kinds of goods.

The active agents that instigate change are as many and as complex as the social system itself. Many so-called causes operate in the same way as do their apparent opposites. Price increase brought increased rice in 1919, but price decrease did much the same thing in 1934. In addition the assigned cause of one effect sometimes produces elsewhere the inverse effect: Decreasing harvests of rice during the 1960s provoked Bang Chan to abandon rice cultivation, but neighboring areas, where new jobs were less accessible, were constrained to redouble their efforts. So if rules exist for social change, we must seek them in a broader context, for the development of a social group seems not to unfold in uniform stages like the development of a fetus. The potentials for development have been overly narrowed; the female presumed to bear only human beings seems also able to bear fish, monkeys, and hitherto undescribed creatures.

CONTINGENCIES. Five thousand years or more may have intervened in China between shifting cultivation of rice, probably discovered by the neolithic Lung Shan people, and the development of transplanting around the start of the Christian era, perhaps about 200 B.C. during the beginnings of imperial rule under the Han dynasty. Eighty-five years sufficed for Bang Chan to reach the same mode of cultivation, and today with a bulldozer, we might have irrigation ditches bringing water to dyked fields within a few months. Where does this compression, this saving of time come from?

Bang Chan had less to learn, less new experience to shape. No one had to discover that rice will tolerate water in depth or that water in sufficient quantities at proper times can reduce hours of weeding. No one had to discover through trial and error the proper distance to set transplanted seedlings and the optimum depth of water for the next days. We do not know how many planting seasons in China began and failed, how many communities of rice growers turned from rice to other crops, or how many were somehow forced to continue in desperation.

Bang Chan not only could take advantage of the cumulated inventions but of local conditions that overrode certain steps in development and compressed others. Mau An and his group could take advantage of natural flooding and begin with broadcasting rather than with shifting cultivation. The already-finished canal system together with natural flooding made conversion to transplanting a relatively simple step, in contrast to locations where long ditches had to be dug to bring the water and where fields required terracing. Unlike the vast plains of western U.S. or the even vaster ones of eastern Siberia, problems of desperately low manpower were relatively short-lived. Bangkok, always lying just over the horizon, dispatched a growing stream of migrants. After its first decades Bang Chan had few vacant areas unused for cultivation or pasture, nor were crops left in the field to rot because they lacked the workers to harvest them.

If our statistics on population density associated with each mode of cultivation are approximately correct (see Table 4.4), a change from broadcasting to transplanting does not ordinarily occur until density reaches 500 per square mile. Bang Chan was probably well below this level when it made the change, possibly about 400 per square mile. Such a change could happen because the needed energy input was less exclusively dependent on human labor than in China, Java, and Vietnam. A lighter population supplied with more draft animals, tools, and machinery could in less than a century match earlier developments requiring many centuries of population increase. Today with a bulldozer and other earth-moving equipment the time is even shorter.

To this we quickly add that development must still be phrased in terms of the feasibility of supplying a given quantum of energy to the making of a certain holding. Advantages are reckoned against disadvantages in whatever accounting system may be current. The Great Wall of China was feasible by balancing the deaths of ten thousand workers against the loss of a country to invaders. Chaem and his brothers, however, felt that a buffalo and plow would make more trouble than preparing the rice plot by hand. He and his brothers knew about buffaloes and plows as surely as their children knew about transplanting, but each generation preferred to let its successor stretch itself to adopt the more demanding technique. Each generation depended less exclusively on the raw human energy available in the locality and more on the energy of other communities that becomes available through machines and markets.

In addition to peculiarities of place which give Bang Chan its own rate of development, there are peculiarities of epoch. Had Bang Chan's appearance failed to coincide with, among other things, the period when world markets and cheap ocean transport were present, its inhabitants might still be happily eating most of the rice that their small plots produced. The contingencies on which Bang Chan's succession of holdings depended or depend thus resemble those on which rice cultivation itself rests. Some are clearly of the same cyclic nature. The Chao Phraya had not cut its banks too deep, nor the mountains decayed to the point where heavy silts make it a river that changes course wildly, such as the Hwang Ho in China or the levied Mississippi. Other contingencies may be cyclic or linear, depending on one's view of history. Is there but one Bangkok that gives its population to the surrounding area, or may we say that many Bang Chans have appeared in the supporting regions of Southeast Asian cities. Some day will Bangkok too become a ruined Pagan or an Angkor Wat? Bang Chan altered its holdings fairly rapidly, perhaps because it arose in a period of peaceful population increase during the rise of a powerful dynasty of Thai kings, but dynasties, population growth, and peace are also exhaustible.

SOME IMPLICATIONS OF A BALANCED HOLDING

Having invoked the concept of ecological holding and optimal output coming from an optimal input, let us examine briefly certain implications of this view for society and man. According to these working assumptions, social problems are to be understood basically as imbalance caused by insufficient input in relation to output or excessive input in relation to output. In both types of social problem we may expect to find more or less chronic shortages, perhaps of food, labor, land, or tools.

To suggest the differences between a social group with insufficient, and one with excess, input for its holding, we have constructed Table 8.1, positing six points of differentiation.

Table 8.1. Schematic differentiation of social groups with insufficient and with excess input in relation to holdings of approximately equal output

	Insufficient Input	Excess Input
1. Division of labor as a whole	flexible	rigid
2. Degree of specialization of labor	low	high
3. Social differentiation as a whole	slight	considerable
4. Population movement	high	low
5. Social mobility	high	low
6. Stability of wealth	low	high

This table outlines our proposition that insufficient and excess input are reflected in certain aspects of social organization. Let us illustrate the points by comparing Alaska, presumed to have insufficient input for its holding, with Puerto Rico, presumed to have excess input for its holding. Though the holdings are not identical, they will serve for illustration. Alaska's division of labor by sex, age, or specialty is less firm, the penalties of breaching norms less severe, than in Puerto Rico. Similarly Alaska has fewer full-time specialists, for one may hunt, fish, garden, and sell insurance to earn a living, according to circumstance. Perhaps this is a way of saying that in Alaska, the groups of people are less crossed by distinctions of class, wealth, occupation, and so forth, than in Puerto Rico, where the emblems of dress, residence, diet, and etiquette are deemed important. While Alaskans are on the move geographically and socially, we may expect to find more Puerto Ricans born, living, and dying in the same village or neighborhood. The wealth of these people varies little in comparison with the fortunes made and lost in Alaska. These distinctions may well resemble more than superficially the differences between a frontier and an urban community, between village and city neighborhood, between young and old settlements.

Bang Chan appears to be an area of insufficient input, especially since embracing the market economy in 1920. Men, women, and children flexibly work where the job is. Until recently all were part-time rice growers, though peddlers, laborers, storekeepers, and government officials have recently entered the scene. However, no more than a generation lies between a rice grower and his son or daughter who is a government official, and often people change occupation from season to season. Despite hierarchic organization of society, high and low, rich and poor once lived and dressed in the same manner. Only the 1960s introduced clear distinctions between the local rice-growing population and the newly arrived city folk with suburban residences. The case for decreasing geographical and social mobility is less clear, though we should expect indices soon to show diminishing migration and variation of status as well as variation of wealth.

When imbalance reaches pathological proportions, poverty results. Someday we may be able to distinguish wherever it may be found the poverty of declining incentives and skills characteristic of an Appalachia with insufficient input from the urban slums with their excess input. Now, however, let us indicate the general remedies for less drastic imbalance.

For insufficient input the temptation lies in reaching for greater mechanization in order to "save labor," though in fact mechanization alleviates the symptom while aggravating the cause. Indiscriminately used, it draws labor from low to high energy input enterprises with resulting declines in production. As workers are lured from agriculture to manufacture by higher wages and shorter working hours, farm after farm gives up production because of insufficient labor and high cost of machines. Consequently food costs rise enormously, as low-energy demanding farms are replaced by food factories. Over the past half century our food factories have only survived with imported labor and costly machinery for producing selected crops that lend themselves to mass handling and long-distance transport, though quality of food has declined when compared with more carefully tended, home-grown produce.

The remedy for insufficient input takes two directions: First, enterprises with low input demands must be conserved rather than displaced. Animals must be retained, machinery avoided wherever possible; distribution of produce on local rather than national levels should be encouraged; "farmers' markets" must be shown preference over retail stores. Thus imbalance can be compensated rather than intensified. Second, production with high-energy requirements must be permitted only selectively after insuring that shortages will not develop or can be tolerated briefly. In the U.S., for example, the space program, the supersonic jets, and comparable programs would have to be studied for their effects on production before they were undertaken.

To reduce excess input by a group requires a more demanding holding. Modern technology is an obvious remedy, though we may expect a concomitant reorganization of tasks at many points to fit the peculiar distinctions and rankings of other cultures. A more difficult problem lies in reallocating workers who have been oriented toward skills and specializations that may no longer be useful. India, for instance, resists mechanical pumps, because they threaten to displace the occupationally specialized drawers of water, yet were it possible for these specialists to manufacture, distribute, sell, maintain, and man these pumps, two or three times the number of people now engaged in drawing water would be more rewardingly employed. Balance is to be achieved by redefining and reallocating tasks, which in India are matters of religion and morality as well as occupational habits.

The accounting system for making these economizing decisions has been devised neither by socialists, capitalists, nor anyone else. Standard accounting systems deal with problems such as costs of materials, capital, and labor for state public works as well as private corporate enterprise. Wage earners then become an expensive or inexpensive ingredient. A wage of $50,000 may be paid to one man for a year's work, or to five or ten workers for the same period; all is the same for the enterprise, but not for group living. In this case the number employed and their allocation is the important criterion, and we have seldom made social decisions with man or men as the center. Reduced to man-days of work per year, few people are five or ten times as valuable as others. A new accounting system would have to be devised that weighs the well-being of men in terms of the man-days of work and its allocation. The complexity of such a system, requiring attention over the life span of a human being, might tax the resources of modern computers.

In such a vista, technology returns to serving as a tool for man.

Only the illusions need be dispelled, such as the promise of extending control over nature. Now it is clear that nature itself is no longer definable in the traditional Greek manner, for little lies outside the influence of man. The last vestiges of virgin forest differ little from second and third growth. The familiar rice field exemplifies again how firmly man's hand has molded a landscape into forms of considerable charm. Yet when man prides himself on the electric illumination that makes him master over nature, his constricted vision overlooks his greater dependencies on rain to maintain the volume of water, on the machines of the hydroelectric plants, and on the slender wires that bring the power to him. If he has gained control over light and dark, he has sacrificed autonomy in doing so.

In less technologically centered times we could see more readily that information, machines, and resources are but one of many activators of societal life. Building bridges, airplanes, research laboratories, hospitals, and government office buildings represent promises of advancing comfort, freedom from distress, greater leisure in a world activated by utopian technology. Depending on the prophets one may choose, its form may be either socialist or capitalist, but the peace described as the reward of our toils is equally inviting. As an activator of higher input in more complex holdings, technology stands as just one among many in the course of human history. Christianity has done as much and, in consideration of its considerably longer duration, a good deal more. Hundreds of cities and buildings have been built in the name of Christ. The towers of Chartres cathedral or the dome of St. Peter's in Rome may not stand as high as the Empire State Building in New York City, but they produced greater riches for the soul in sculpture, painting, and music. Because of Christianity, people made pilgrimages and crusades, if not to the moon, to remote and no less dangerous places to spread the gospel. People have taken up arms for Christianity as well as to gain control over strategic resources. The terror of hell and judgment day was no less acute than the fear of extermination by nuclear bombs dropped by foreign devils, and the promise of a world of leisured plenty has its parallel in Christian heaven. Islam has done no less in changing the holdings of people.

Whether ergs activate ideas or ideas activate ergs goes beyond the scope of this study, yet clearly ideas as well as physically measurable power has activated human society throughout the ramblings of history. Just what led certain people of the Fertile Crescent to build the temples and walls of Sumer and Ur is known only in a general way, but that activation was total, filled with terrors and promises, engineering feats, music, and morally proper organization of people. The demon that rode on the shoulder of Genghis Khan whispered different instructions and produced different effects than the books of Marx and Lenin indicating the way to the leaders of an expanding Russia, but the activations shook the earth no less. In our day lesser people erect buildings, take long trips, and utter prophetic statements in the name of alcohol, drugs, tobacco, clear streams, and dumb animals.

Countries that we now scale from developed to underdeveloped on the basis of technological accomplishments may be seen in fresh perspective when viewed as societies with their peculiar holdings. Though some have failed to respond to the promises and terrors of technology, they may so respond in the future, or they may be of a type that is unsusceptible to technological activation. Religion, art, or conversation may serve them better.

Finally a word about morals, for every thesis about society carries a manifest or covert guide to action. The notion of input increases leading to greater output offers a morality of energy, a Nietzschean morality of the strong, but the concept of holding modifies this position with the theme of balance. The optimal return comes not from the total quantity of input but from the proper amount in accordance with the requirements of the holding. Skill in remodeling a holding to fit one's capacity to supply input, skill in adaptation, these are the needed qualities together with awareness, sensitivity, and perceptiveness of the broad environmental arc. One man can perhaps reap the rice from a plot of land in one day by working hard with his sickle, though most others can reap but half that amount. The niceties lie not in producing an input of this size, but in reckoning whether one can sustain this double rate along with a hundred other obligations.

The ideal man is the person of many sensitivities and multiple dexterities. He is neither the scholar specialist nor the warrior, neither the artist nor the scientist, neither the toiler in the field nor the languid observer in the reclining chair, but a combination of them all. His sensitivity radiates through a broad arc, and his energies still remain ready to meet the next demand. He is the discriminating child who cares for aging parents, never doing enough to threaten their feelings of adequacy, always enough to reassure them of his devotion. He can smile in passing a friend, never so slightly as to appear condescending, nor so effusively as to appear insincere. Like a chamber-music player he hears both his own instrument and those of the other members of the group, plays for the totality rather than to be heard above others, and compensates adroitly for the limitations of his instrument and his own skill. No task exists that cannot lure his talents, but he participates with precise awareness of his shortcomings.

Evil in the world is not killing and destruction per se. Death is an aspect of living, like birth and puberty. As implied in the concept of "food chain," one species depends upon the death of another for its survival. Indeed, for human beings the death of rice is its most significant moment, as if its whole existence derives meaning only when the fruit has been broken from the stalk and beaten to inertness. Men as well as maggots and mold live on death.

Killing and destroying are sins in a limited sense. They are the least suitable responses for most occasions, since inevitably the victim resists, the killer is repulsed by his own action, and the social group is left with a raw edge. They are devices to be used only when no alternative exists; a person of sensitivity and dexterity would avoid them as long as possible.

Evil is the energy input unbalanced by the requirements of a holding. We have seen this evil among the rice growers who plow and sow but fail to weed sufficiently and harvest carelessly. It reappears in the rice growers of the uplands who fail to allow the soil to regenerate before planting a new crop. Elsewhere social leeches take from the holding but fail to make their contributions; they are the idle rich and the idle poor. Evil is the blindness of persons eminently trained in their specialties but ignorant of their limitations. They are the makers of indestructible nerve gas, nuclear warheads, and undecomposing pesticides. These cases of inadequate input may be matched by the excesses of others. Here among others is the evil of waste—the resources of material and human labor going into products that will be outmoded rather than outworn, the weapons made at a cost of millions of man-days to kill a dozen persons in a distant jungle.

This and more is clearly evil as soon as we have recognized the many-faceted dependencies on which human society rests. We require not only other human beings, domesticated animals, fresh air and water, but insects, birds, and fish to sustain us. We also depend on the longer life cycles of matter, the formation of soil, and the formation of minerals that may require millions of years of cosmic time. In this sense the world is sacred, and we are obliged to treat it in that way. Productive toil is our sacred obligation.

Appendix A

ORTHOGRAPHY AND PRONUNCIATION OF THAI WORDS

Inevitably we have had to use Thai words to describe Thailand, and many of them, such as the names of cities and rivers, are familiar enough—Mekong, Chao Phraya, Bangkok. Others lack a standard English spelling, such as the names of persons—Chaem, Bun, Mau An, Phlym. We have written such words using basically the orthography of Dr. Mary Haas stripped of diacritical marks for tone and vowel length. A rough approximation of the Thai sounds then results by observing the following rules:

1. All consonants are to be treated as initial consonants in English words except for the following:

    k unaspirated as the final *ck* in *brick;*
    kh aspirated as the initial *k* in *king;*
    t unaspirated as the final *t* in *pot;*
    th aspirated as the initial *t* in *tin;*
    p unaspirated as the final *p* in *tip;*
    ph aspirated as the initial *p* in *pit;*
    ng as the final *ng* in *gong;*
    c as the initial *j* in *jug.*

2. All vowels are to be treated as in Latin and Italian except for the following:

    ae as *a* in *hat;*
    au as *au* in *auto;*
    aw as *ow* in *how;*
    y as *eu* in the French word *feu.*

Appendix B

SOURCES OF NUMERICAL DATA ON RICE PRODUCTION

Sources for Table 4.2. Modes of cultivation and their yields of rice

Location	Yield (short tons/acre)	Source
Shifting Cultivation		
Northwestern Laos	1,470	Izikowitz, 1951, p. 287
Java	825	von Bernegg per Izikowitz, 1951, p. 287
Northern Thailand	2,246	Miles, n.d.
Central Vietnam	2,530	Lafont, 1951 per Halpern, 1958, p. 32
Burma, Malaya	1,680	Jack, Craig, Leach per Freeman, 1955, p. 98
Sarawak	917	Freeman, 1955, p. 98
Broadcasting		
Cambodia (floating rice)	880–2,200	Delvert, 1961, pp. 674–75
Northern Thailand (Chiengrai)	1,870–2,860	Moerman, 1968, p. 162
Central Thailand (Ayuthia, Lopburi)	990–1,760	Zimmerman, 1931, p. 33
Burma (Akyab)	1,650–1,870	Nuttonson, 1963, Tables 50, 51
Transplanting		
Cambodia (Siem Réap, Takeo)	550–2,420	Delvert, 1961, pp. 674–75
Northern Thailand (Chiengrai)	2,530–2,970	Moerman, 1968, p. 162
Central Thailand (Bang Chan)	2,178	Janlekha, 1955, p. 52
China (Yunnan)	550–1,760	Chen, 1949, p. 42
Northern Thailand (as whole)	2,612	Chapman, 1967, Table 3
North Vietnam	2,968	Gourou, 1936, p. 393
Philippines (Luzon)	2,320–5,170	IRRI, 1967

Sources for Table 4.3. Modes of cultivation and variability of yield over successive years in the same field

Location	Variability*	Years of Observation	Source
Shifting Cultivation			
North Thailand (Chiengrai)	54.4	3	Hanks, Sharp, Hanks, 1968, p. 29
Broadcasting			
North Thailand (Chiengrai)	21.0	2	Moerman, 1968, p. 160
Transplanting			
North Thailand	10–14	2	Moerman, 1968, p. 160
Central Thailand	10–22	4	Janlekha, 1955, p. 52

$$\text{* Crop variability} = \frac{\text{Maximum crop} - \text{minimum crop} \times 100}{\text{Maximum crop}}$$

Sources for Table 4.4. Modes of cultivation and their associated population densities

Location	Density (persons/square mile)	Source
Shifting cultivation		
Laos (Lamet area)	7.5	Izikowitz, 1951, p. 38
Laos (Nam Tha)	13.0	Halpern, 1961, Table 6
Thailand (Chiengrai)	23.8	Hanks, Sharp, Hanks, 1968, p. 5
Sarawak	52–65	Freeman, 1955, p. 133
Philippines (Mindoro)	65–91	Conklin, 1957, p. 10
Indochina	13	Gourou, 1951, p. 31
Broadcasting		
Cambodia (Battembang)	47–465	Delvert, 1961, p. 640
Northern Thailand	455	Moerman, 1968, pp. 8, 36
Transplanting		
Cambodia (Koh Samrong, Lovea, Siam Réap)	260–1,300	Delvert, 1961, pp. 308, 628
Central Thailand	450	Janlekha, 1955, pp. 25, 48
Philippines (Palawan)	560	Pelzer, 1945, p. 84
North Vietnam	1,120	Gourou, 1936, p. 118

Source for Table 4.5. Modes of cultivation and their human labor requirements per unit of land per crop season

Location	Labor Requirement (man-days/acre)	Source
Shifting cultivation		
Sarawak	102, 42	Geddes, 1954, p. 68
Sarawak	50–71	Freeman, 1955, p. 90
Broadcasting		
Cambodia	17–24 (estimate)	Delvert, 1961, pp. 347–48
Northern Thailand	14.3	Moerman, 1968, p. 205
Transplanting		
Cambodia	24–33	Delvert, 1961, pp. 347–48
Northern Thailand	37.4	Moerman, 1968, p. 203
Central Thailand	36.2	Janlekha, 1955, p. 106
Sarawak	135, 92	Geddes, 1954, p. 68
North Vietnam	90	Dumont, 1954, p. 138
Taiwan	30	Alicbusan, 1964
Taiwan	70	Chang, 1963
Thailand	38–35	IRRI, 1966, p. 243
Philippines	34–30	IRRI, 1966, p. 243

Sources for Table 4.6. Mode of cultivation and size of cultivated field worked by a single household

Location	Field Size (acres)	Source
Shifting Cultivation		
Northwestern Laos	1–7	Izikowitz, 1951, p. 258
Thailand (Tak)	2–3.25	Moorman, 1964, p. 6
Thailand (Chiengrai)	6	Miles, n.d., p. 9
Sarawak	4	Freeman, 1955, p. 93
Burma	2.5–3	Leach per Freeman, 1955, p. 93
Indonesia	2.5	Pelzer per Freeman, 1955, p. 93
Broadcasting		
Cambodia (Battembang)	11.5–5.8	Delvert, 1961, pp. 473, 642
Central Thailand (Lopburi, Ayuthia)	12–10	Zimmerman, 1931, p. 33
Transplanting		
Cambodia (Kampong Cham, Prey Veng, Suaj Rieng)	4.3–10	Delvert, 1961, p. 472
Central Thailand (Bang Chan)	12	Janlekha, 1955, p. 58
Central Thailand (Prakanong)	6	Zimmerman, 1931, p. 33
China (Yunnan)	4.3	Chen, 1949, p. 37
Northern Thailand	2.5	Chapman, 1968, Table 1
North Vietnam	1.3–2.5	Dumont, 1954, p. 135

BIBLIOGRAPHY OF REFERENCES IN TABLES

References cited in the text are marked by an asterisk.

Alicbusan, L. C. "The Rate of Substitution of Man Hours by Animal and Machine Horsepower Hours in Rice Production." Mimeographed. Armidale, New South Wales: University of New England Department of Geography.

Chang Cheu-Chang. "An Agricultural Engineering Analysis of Rice Farming Methods in Taiwan." Farm Machinery Research Center, National Taiwan University, AE-M Report No. 5, 1963.

Chapman, Theodore. "An Appraisal of Recent Agricultural Changes in the Northern Valleys of Thailand." mimeo. 1967.

*Chen Han-seng. *Frontier Land Systems in Southernmost China.* New York: International Secretariat, Institute of Pacific Relations, 1949.

*Conklin, Harold C. "Hanunoo Agriculture: A Report on an Integral System of Shifting Cultivation in the Philippines." FAO Forestry Development Paper No. 12. Rome: Food and Agriculture Organization of the United Nations, 1957.

Delvert, Jean. *Le Paysan Cambodgien.* Paris: Mouton, 1961.

Dumont, René. *Types of Rural Economy: Studies in World Agriculture.* London: Methuen, 1954.

*Fei, Hsiato-tung and Chih-i Chang. *Earthbound China: A Study of Rural Economy in Yunnan.*" London: Routledge 1948.

Freeman, J. D. *Iban Agriculture.* London: Her Majesty's Stationery Office, 1955.

*Geddes, W. R. *The Land Dyaks of Sarawak.* London: Her Majesty's Stationery Office, 1954.

Goldsen, Rose K., and Max Ralis. "Factors Related to Acceptance of Innovations in Bang Chan, Thailand." Data Paper No. 25, Southeast Asia Program, Department of Far Eastern Studies, Cornell University, Ithaca, N.Y., 1957.

Gourou, Pierre. *Les Paysans du Delta Tonkinois.* Paris: Les Editions d'Art et d'Histoire, 1936.

Gourou, Pierre. "Utilization of Upland Areas of Indochina." Part II in *Development of Upland Areas in the Far East.* New York: Institute of Pacific Relations, 1951.

Halpern, Joel. "Population Statistics and Associated Data." Laos Project Paper No. 3. mimeo. 1961.

Hanks, Lucien M., Lauriston Sharp, and Jane R. Hanks. *A Summary of Data from the Mae Kok Area of Thailand with some Afterthoughts on Upland Settlement Patterns.* unpublished, 1968.

IRRI (International Rice Research Institute). *Agricultural Economics Annual Report for 1967*. Los Banos, Philippines. mimeo. 1967.

Izikowitz, Karl G. "Lamet: Hill Peasants in French Indochina." Ethnologiska Studier No. 17. Goteborg: Ethnografiska Museet, 1951.

*Janlekha, Kamol. "A Study of the Economy of a Rice Growing Village in Central Thailand." Bangkok: Ministry of Agriculture, Division of Agricultural Economics, 1955.

Miles, Douglas. "Shifting Cultivation: Threats and Prospects." Tribal Research Center, University of Chiengmai (Thailand). mimeo. n.d.

*Moerman, Michael. *Agricultural Change and Peasant Choice in a Thai Village*. Berkeley: University of California Press, 1968.

Moorman, F. R., K. R. M. Anthony, and Samarn Panichapong. "Note on the Soils and Land Use in the Hills of Tak Province." Bangkok: Ministry of National Development, 1964.

Nuttonson, M. Y. "The Physical Environment and Agriculture of Burma." Washington, D.C.: American Institute of Crop Ecology, 1963.

Pelzer, Karl J. *Pioneer Settlement in the Asiatic Tropics*. New York: American Geographical Society, 1945.

Zimmerman, Carle C. *Siam Rural Economic Survey 1930–31*. Bangkok: The Bangkok Times Press, 1931.

SUGGESTED FURTHER READING

References cited in this volume are marked by an asterisk.

Attagara, Kingkeo. *"The Folk Religion of Ban Nai."* Ph.D. thesis. University of Indiana, 1967. A community study in Thailand with special focus on the folklore and ceremonial life.

Bates, Marston. *Where Winter Never Comes: A Study of Man and Nature in the Tropics.* New York: Scribner's, 1952. An ecologist stresses the peculiarities of living in a tropical setting.

*Boserup, Ester. *The Conditions of Agricultural Growth.* Chicago: Aldine, 1965. An economist approaches the conditions of technological change in agriculture as the conditions for utilization of labor.

*Burkhill, L. H. Dictionary of the Economics Products of the Malay Peninsula. London: Government of the Straits Settlements and Federated Malay States, 1935.

*Burma Research Society Fiftieth Anniversary Publications, No. 2. Rangoon: Burma Research Society, 1959.

*Condominas, Georges. *Nous Avons Mangé la Forêt.* Social and ceremonial complexities in the agricultural cycle of shifting cultivators in the highlands of Vietnam. Paris: Mercure de France, 1957.

*Fei, Hsiao-tung and Chih-I Chang. *Earthbound China: A Study of Rural Economy in Yunnan.* London: Routledge, 1948. Part I describes intensive rice farming of small holdings in the context of Chinese village life.

Geertz, Clifford. *Agricultural Involution: The Processes of Ecological Change in Indonesia.* Berkeley: University of California Press, 1963. An outline of social changes from shifting cultivation of rice to transplanting.

*Gervaise, Nicholas. *The Natural and Political History of the Kingdom of Siam, A.D. 1688.* Trans. by Herbert Stanley O'Neill, Bangkok, 1928.

*Grist, D. H. *Rice.* London: Longmans, 1959. A standard reference on the characteristics of rice, its cultivation and processing.

Hanks, Jane Richardson. "Reflections on the Ontology of Rice," *Primitive Views of the World,* Stanley Diamond, ed. New York: Columbia University Press, 1960.

Izikowitz, Karl G. "Lamet: Hill Peasants in French Indochina." Ethnologiska Studier No. 17. Goteborg: Ethnografiska Museet, 1951. Life among shifting cultivators in the hills of Laos.

Kaufman, Howard K. *Bangkhuad: A Community Study in Thailand.* Locust Valley, N.Y.: J. J. Augustin, 1960. Description of a rice-growing village in the neighborhood of Bang Chan.

*King, D. O. "Travels in Siam and Cambodia." Journal of the Royal Geographic Society 30, 177–82. 1860.

*Legge, James, trans. *The Texts of Taoism.* London: Muston, 1927.

*Matsuo, Takane. *Rice and Rice Culture in Japan.* Tokyo: Institute of Asian Economic Affairs, 1961. A technical study of rice with details of its cultivation in a country of advanced methods.

*Moerman, Michael. *Agricultural Change and Peasant Choice in a Thai Village.* Berkeley: University of California Press, 1968. Rice growers in northern Thailand and their range of modes of cultivation.

*Murdock, G. P. *Social Structure in Southeast Asia.* Chicago: University of Chicago Press, 1960.

*Phillips, Herbert P. *Thai Peasant Personality: The Patterning of Interpersonal Behavior in the Village of Bang Chan.* Berkeley: University of California Press, 1965. Aspirations and attitudes of persons working in the rice fields.

*Rawson, R. R. *The Monsoon Lands of Asia.* Chicago: Aldine, 1963. The economic geography of India and Southeast Asia.

Spencer, J. E. *Shifting Cultivation in Southeastern Asia.* Berkeley: University of California Press, 1966. A geographic approach to the conditions of shifting cultivation.

*Waley, Arthur, trans. *The Book of Songs.* London: George Allen & Unwin, Ltd., 1954.

The cover photograph and the first, third, and twelfth photos in the plate section are by Ewing Krainin. All others were provided by Jane R. Hanks.

Index

ments that facilitate their work, Mr. Hanks shows how their lives changed as they adopted each new mode of cultivation. By drawing together what contemporary residents say about themselves and their grandparents and fresh observations of still existing pioneer settlements in other parts of Thailand, the author fashions a sequence of ecological portraits of the four periods comprising the brief history of Bang Chan. Concise and lucid, this book is not only a data-rich analysis but also a human portrayal of the ecosystem of rice and man in Southeast Asia.

About the Author

Lucien M. Hanks has taught at the University of Illinois, Bennington College, the University of California, and the University of Vermont. He is the author of numerous articles in anthropology and is currently Senior Research Associate in the Southeast Asia Program, Cornell University.